T0077235

Cyber Security Applications for Industry 4.0

Cyber Security Applications for Industry 4.0 (CSAI 4.0) provides integrated features of various disciplines in Computer Science, Mechanical, Electrical, and Electronics Engineering defined as Smart systems. It is eminent that Cyber-Physical Systems (CPS) provide accurate, real-time monitoring and control for smart applications and services. With better access to information from real-time manufacturing systems in industrial sectors, the CPS aim to increase the overall equipment effectiveness, reduce costs, and improve efficiency. Industry 4.0 technologies are already enabling numerous applications in a variety of industries. Nonetheless, legacy systems and inherent vulnerabilities in an organization's technology, including limited security mechanisms and logs, make a move to smart systems particularly challenging.

Features:

- Proposes a conceptual framework for Industry 4.0-based Cyber Security Applications concerning the implementation aspect
- Creates new business models for Industrialists on Control Systems and provides productive workforce transformation
- Outlines the potential development and organization of Data Protection based on strategies of cybersecurity features and planning to work in the new area of Industry 4.0
- Addresses the protection of plants from frost and insects, automatic hydroponic irrigation techniques, smart industrial farming, and crop management in agriculture relating to data security initiatives

The book is primarily aimed at industry professionals, academicians, and researchers for a better understanding of the secure data transition between the Industry 4.0-enabled connected systems and their limitations.

Chapman & Hall/CRC Cyber-Physical Systems

Series Editors:
Jyotir Moy Chatterjee
Lord Buddha Education Foundation, Kathmandu, Nepal
Vishal Jain
Sharda University, Greater Noida, India

Cyber-Physical Systems: A Comprehensive Guide
By: Nonita Sharma, L K Awasthi, Monika Mangla, K P Sharma, and Rohit Kumar

Introduction to the Cyber Ranges
By: Bishwajeet Pandey and Shabeer Ahmad

Security Analytics: A Data Centric Approach to Information Security
By: Mehak Khurana and Shilpa Mahajan

Security and Resilience of Cyber Physical Systems
By: Krishan Kumar, Sunny Behal, Abhinav Bhandari, and Sajal Bhatia

Cyber Security Applications for Industry 4.0
By: R Sujatha, G Prakash, and Noor Zaman Jhanjhi

For more information on this series please visit: https://www.routledge.com/Chapman--HallCRC-Cyber-Physical-Systems/book-series/CHCPS?pd=published,forthcoming&pg=1&pp=12&so=pub&view=list?pd=published,forthcoming&pg=1&pp=12&so=pub&view=list

Cyber Security Applications for Industry 4.0

Edited by
R Sujatha
G Prakash
Noor Zaman Jhanjhi

CRC Press is an imprint of the
Taylor & Francis Group, an **informa** business
A CHAPMAN & HALL BOOK

Cover image: Akash Sain/Shutterstock

First edition published 2023
by CRC Press
6000 Broken Sound Parkway NW, Suite 300, Boca Raton, FL 33487-2742

and by CRC Press
4 Park Square, Milton Park, Abingdon, Oxon, OX14 4RN

CRC Press is an imprint of Taylor & Francis Group, LLC

ISBN: 978-1-032-06620-2 (hbk)
ISBN: 978-1-032-06621-9 (pbk)
ISBN: 978-1-003-20308-7 (ebk)

DOI: 10.1201/9781003203087

Typeset in Palatino
by SPi Technologies India Pvt Ltd (Straive)

Contents

Preface

> "I really think that if we change our own approach and thinking about what we have available to us, that is what will unlock our ability to truly excel in security. It's a perspective exercise. What would it look like if abundance were the reality and not resource constraint?"
>
> **-Greg York, VP, Information Security, Tribune Media, at SecureWorld Chicago**

The Fourth Industrial Revolution (Industry 4.0) explores the digitization of Conventional Manufacturing and Control Systems. Industry 4.0 is considered the dawn of "Cyber-Physical Systems (CPS)" which entirely involves innovative competencies for the researchers and people involved. Today, the CPS and IoT devices deployed in the Industrial Control Systems are the blended machines that operate in conjunction with the legacy instruments, which lack sufficient security controls over data manipulation. The risks and threats associated with the Industry 4.0 systems will create an avenue for potential attack surfaces and vectors. This book series aims to provide a detailed review, analysis, and investigation of CPS and the security vulnerabilities that lead to the new path for both academic and professional researchers.

The Editors of Cyber Security Applications for Industry 4.0 (CSAI 4.0) intended to publish this book for the integrated engineering disciplines of Computer Science, Mechanical, Electrical, and Electronics Engineering. It covers the conceptual view of CPS, including current technological revolutions, challenges, and the fusion of CPS with IoT and Energy systems. This book's chapters will bring together the various sectors of industry. The amalgamating structure of CPS contributes to the academia and education sector to open up seamless practical solutions for Healthcare, Agriculture, and Power Grid applications. We received 20 chapters from different authors creating an enduring techno-platform between academia, industry, and research groups. After meticulously scrutinizing all the submissions, ten chapters have been selected and considered for publishing.

The book is structured into ten chapters, with the following primary features:

- Coverage of principles of the Cyber-Physical Systems
- Discussions on recent developments in Industry 4.0
- Deliberation of technological advancements in diversified domains
- Adequate content for Undergraduate, Postgraduate, and Ph.D. students who aim for comprehensive knowledge in CPS.

The Publication Partner for CSAI 4.0 is an eminent American publishing group – CRC Press – that specializes in producing technical books. CRC Press is now under the aegis of renowned Routledge Taylor and Francis Group. We want to express our heartfelt gratitude to the Publishing team for their invaluable assistance and guidance throughout the process of getting our book chapters published swiftly. We thank all the authors for submitting the good-quality book chapters and express our humble gratefulness to various experts who helped us in the review process with valuable suggestions. Lastly, we appreciate the great teamwork put forward to host this book on CSAI 4.0.

R Sujatha
G Prakash
Noor Zamam Jhanjhi

About the Editors

Prof. Dr. R. Sujatha completed her Ph.D. degree at Vellore Institute of Technology, Vellore, India, in 2017, in the area of data mining. She received her M.E. degree in Computer Science from Anna University, India, in 2009 with university ninth rank and did a Master of Financial Management from Pondicherry University, India, in 2005. She received her B.E. degree in Computer Science from Madras University, India, in 2001. She has 17 years of teaching experience and has been serving as an associate professor in the School of Information Technology and Engineering in Vellore Institute of Technology, Vellore. She has organized and attended a number of workshops and faculty development programs. She actively involves herself in the growth of the institute by contributing to various committees at both academic and administrative levels. She used to guide projects for undergraduate and postgraduate students and currently guides doctoral students. She gives technical talks in colleges for symposiums and various sessions. She acts as an advisory, editorial member, and technical committee member in conferences conducted in other educational institutions and in-house too. She has published a book titled *Software Project Management for College Students*. She has also published research articles in reputed high-impact journals. The Institution of Green Engineers awarded her IGEN Women Achiever 2021 award in the Future Computing category. She is interested in exploring different places and visiting them to know about the culture and people of various areas. She is interested in learning upcoming things and gets herself acquainted with the student's level. Her areas of research interest include Information Management and Data Analytics, Artificial Intelligence, Soft Computing, Medical AI, Big Data, Deep Learning, and Blockchain Software Engineering

Dr. G. Prakash is working as an Assistant Professor (Selection Grade) in the Department of Computer Science and Engineering, Amrita School of Engineering, Amrita Vishwa Vidyapeetham University, Bengaluru campus, India. He is currently guiding six Ph.D. scholars and two postgraduate scholars specifically in the domain area of Cyber Security. He served as the Session Chair for the Special sessions of ISIC 2021 and

ISIC 2022 on the following titles: *Smart Knowledge Acquisition through Semantic Intelligence* (SKASI21); *Cyber-Physical Big Data Systems* (CPBDS 2022) organized by MERI College, New Delhi, India, in February 2021 and by Georgia Southern University (Armstrong Campus), Savannah, the United States, in May 2022. He also served as the Session Chair for the Special Track of CIS 2020 on the title *Fusion of Artificial Intelligence and Blockchain Technologies* (FAIBT 2020) organized by Soft Computing Research Society, New Delhi, in September 2020. He is the Discussant and TPC member for the 11th and 12th EAI International Conference on Digital Forensics & Cyber Crime (EAI ICDF2C 2020 AND 2021, Singapore. He was designated as a Conference Chair for the International Conference on Intelligent Computing (ICIC 2018) and Special Editor of the Scopus Indexed Journal – *Journal of Computational and Theoretical Nanoscience* – where 78 conference papers were presented and published. He has done Summer Research Fellowship at the Indian Institute of Science, Bengaluru, India. He has completed two Sabbatical projects at M/s. Infosys Technologies Limited, Bengaluru, in the relevant area of Network Security. He has delivered a Technical Talk at the two-day International Virtual Symposium organized by the University of Missouri and Amrita Vishwa Vidyapeetham under the Joint Research Center in Cyber Physical Systems in February 2021.

Prof. Dr. Noor Zaman Jhanjhi (NZ Jhanjhi) is currently working as Associate Professor, Director Center for Smart society 5.0 [CSS5], and Cluster Head for Cybersecurity cluster, at the School of Computer Science and Engineering, Taylor's University, Malaysia. The cybersecurity research cluster has extensive research collaboration globally with several institutions and professionals. Dr Jhanjhi serves as Associate Editor and Editorial Assistant Board for several reputable journals, such as PeerJ Computer Science, Frontier in Communication and Networks. He received Outstanding Associate Editor for IEEE ACCESS for 2020, PC member for several IEEE conferences worldwide, and guest editor for several reputed indexed journals. Active reviewer for a series of top-tier journals has been awarded globally as a top 1% reviewer by Publons (Web of Science). He has high indexed publications in WoS/ISI/SCI/Scopus, and his collective research Impact factor is more than 500 points. He has several international Patents on his account, including **Australian, German, Japan,** etc. edited/authored more than **40** research books published by world-class publishers, including **Springer, Taylors and Frances, Willeys, Intech Open, IGI Global USA,** etc. He has great experience supervising and co-supervising postgraduate students, and more than **20** PG scholars are graduated under his supervision, and an ample number of current PG students are under his supervision. He is an **external Ph.D./**

Master thesis examiner/evaluator for several universities globally. He has completed more than **30** internationally funded research grants successfully. He has served as a **Keynote/Invited speaker** for more than **30** international conferences globally, presented several Webinars worldwide, **chaired** international conference sessions, and provided Consultancy on several projects internationally. He has vast experience of academic qualifications including **ABET, NCAAA, and NCEAC** for **10** years. His research areas include Cybersecurity, IoT security, Wireless security, Data Science, Software Engineering, and UAVs.
https://expert.taylors.edu.my/cv/noorzaman.jhanjhi

Contributors

D. Aju
Vellore Institute of Technology,
Vellore, India

Imdad Ali Shah
Taylor's University,
Selangor, Malaysia

Fathi Amsaad
Eastern Michigan University,
Ypsilanti, Michigan, USA

Carlos Ankora
Vellore Institute of Technology,
Vellore, India

N. Arulkumar
CHRIST (Deemed to be University),
Bangalore, India

M. Arunadevi
Cambridge Institute of Technology,
Bangalore, India

Agnim Chakraborty
Vellore Institute of Technology,
Vellore, India

Vasavi Chithanuru
Vellore Institute of Technology
Vellore, India

A. Leo Fernandez
Kirirom Institute of Technology,
Traeng, Trayueng, Cambodia

Mamoona Humayun
Al-Jouf University,
Al-Jawf, Saudi Arabia

Noor Zaman Jhanjhi
Taylor's University,
Selangor, Malaysia

M. Kalaiyarasi
V.S.B. Engineering College,
Karur, India

S. Karthi
V.S.B. Engineering College,
Karur, India

Azeem Khan
Taylor's University,
Subang Jaya, Malaysia

M. Kiruthigga
Redfowl Infotech Pvt. Ltd.,
Namakkal, India

M. Lavanya
SASTRA (Deemed to be University),
Thanjavur, India

M. Nandhini
V.S.B. Engineering College,
Karur, India

N. Narmatha
V.S.B. Engineering College,
Karur, India

Adla Padma
Vellore Institute of Technology,
Vellore, India

G. Prakash
Amrita School of Engineering,
Amrita Vishwa Vidyapeetham,
Bengaluru, India

G. Rajarajan
Vellore Institute of Technology
Vellore, India

Mangayarkarasi Ramaiah
Vellore Institute of Technology,
Vellore, India

Vinayakumar Ravi
Prince Mohammad Bin Fahd
University,
Khobar, Saudi Arabia

Abdul Razaque
International Information
Technology University,
Almaty, Kazakhstan

S. Saravanan
Srinivasa Ramanujan Centre
(SRC), SASTRA (Deemed to be
University),
Kumbakonam, India

R. Sujatha
Vellore Institute of Technology,
Vellore, India

Vatsal Vatsyayan
Vellore Institute of Technology,
Vellore, India

Chandra Mouli Venkata Srinivas
BVC Engineering College
East Godavari, India

1

A Detailed Investigation of Popular Attacks on Cyber Physical Systems

Vatsal Vatsyayan, Agnim Chakraborty and G. Rajarajan

Vellore Institute of Technology, Vellore, India

A. Leo Fernandez

Kirirom Institute of Technology, Traeng, Trayueng, Cambodia

CONTENTS

DOI: 10.1201/9781003203087-1

1.1 Introduction: Background and Driving Forces

It has been observed that over the years, there has been a monumental growth in the use of Cyber Physical Systems (CPSs) in various industries including Manufacturing, Energy, Healthcare, and Construction among several others. This can be attributed to the worldwide adoption of Industry 4.0 technologies and the wishes of organizations to move toward a more robust workplace where automation is a priority which would eventually lead to a decrease in costs, higher productivity, and efficiency among other benefits.

The underlying issue that is incurred with this transition is the introduction of several new vulnerabilities, which naturally arise as machines become decentralized and start communicating with each other creating an open environment full of vulnerabilities which can be leveraged by malicious attackers in order to compromise the network and gain control of the entire infrastructure.

In this chapter, we have conducted a literature survey in order to understand the approach that has been taken in the past and explore the underlying problems existing in the industry; then, we have identified the existing threats and significant attacks that have happened in the past so as to gain a proper understanding as to why it occurred and understand how the attacker was able to take advantage of the vulnerability. We have also pointed out the limitations of the CPSs in the industry and finally, we have suggested some future directions in order to prevent attacks and create a more secure environment.

The chapter is distributed into the following sections:

1) Introduction
2) Literature Survey
3) Threats and Popular Attacks
4) Limitations
5) Security Measures
6) Future Direction
7) Conclusion

1.2 Literature Survey

Alguliyev et al. (2018) [1] have identified that CPSs are the groundwork for key industries in the revolutionizing Industry 4.0 such as smart manufacturing, smart city, smart medicine, wearable technologies, etc. They have defined CPS as the integration of physical and computing processes which includes network monitors, embedded computers, and controllers having feedback. The researchers have stated that with the increase in the use of wireless sensor networks over the years, it is paramount to develop information security systems. They have classified CPS threats as Denial of Service (DoS), information disclosure, tampering with data, spoofing identity, and repudiation of origin. They have proposed incorporating authentication mechanisms to CPS components, ensuring methods are in place to protect personnel data and developing countermeasures.

Yaacoub et al. (2020) [2] have identified three central components of CPS as sensors, actuators, and aggregators. They have also emphasized that the use of CPS can lead to increased safety as well as decreased costs. The researchers have also realized several attacks that have happened, for example, physical attacks such as Fake Identity, Physical Breach, and Social Engineering are increasing. Cyber-attacks such as Phishing, Distributed Denial of Service (DDoS), Malware, Botnets, and Watering-hole Attack are on the rise. The researchers have stated that due to the heterogeneous nature of CPS, large-scale deployment, and their dependency on sensitive data, any exposure be it intentional or unintentional can have devastating consequences. They have identified that threats are both physical and cyber. The physical threats include any physical damage that may be inflicted as well as spoofing or sabotaging the system or DoS whereas the cyber threats are unauthorized access, disclosure of information, interception, and reconnaissance. The researchers have identified several vulnerabilities such as network vulnerabilities which create scope for several attacks like DDoS, Man in the Middle attack, replay attacks, sniffing, or spoofing. The cyber vulnerabilities identified were the continual use of open standard protocols, long or short-range wireless communication which creates avenues for different types of attacks, and the most important vulnerability of physical security was the lack thereof, the researchers have identified that there is insufficient physical security to safeguard CPS components. The researchers have gone on to propose making use of Cryptographic-based solutions which go a long way in ensuring the information security triads such as Confidentiality, Integrity, and Availability as well as Authentication. They further suggest the use of non-cryptographic-based solutions such as Firewalls and Intrusion Detection Systems (IDSs). The researchers have suggested enforcing security policy and compliance, defining privileges, regular audits, as well as incident response, periodic employee training, and scheduled risk assessments.

Stefanov and Liu (2014) [3] have outlined immense impacts on the smart grid due to cyber-attacks leading to blackouts, equipment damage, and loss of load. One of the most common vulnerabilities that was found was misconfigured firewalls which create scope for DoS or Intrusion-based attacks. The researchers have analyzed the current CPSs with the help of a testbed and shown that the existing vulnerabilities present in the configurations can be exploited, brute force attacks can be employed to crack the relatively simple passwords that are in place, and the attacker can then take over the system to cause heavy damage. The researchers have recommended advanced firewalls and anomaly-based IDSs to work in synergy.

Wells et al. (2014) [4] have stated that there is a false sense of security in the CPSs owing to the fact that 100% encryption is adopted and the researchers highlight that there is a need for further research on cyber-attacks in order to fully grasp the capabilities and consequences that may arise. They have stated that even simple attacks have the capability of disrupting performance to a significant extent.

Heartfield et al. (2018) [5] have stated that with the increase of smart home technologies in recent times and usage of Internet of Things (IoT) devices employed in households, there is an increase in vulnerability to cyber threats in home, many of which are traditional threats but some are entirely new. They have focused on the impact of CPS threats in smart homes and have identified several threats. Breach of privacy was identified as a physical threat since a lot of data are broadcast at regular intervals, eavesdropping attacks become very dangerous in this scenario, and privacy can be breached through audio/video means, and incorrect actuation has led to community-wide black-outs. The cyber impact observed were threats to confidentiality, integrity by unauthorized manipulation of data, and availability by DoS by jamming communications. The consequences of these attacks can be financially expensive as it creates opportunities for blackmailing victims, installing ransomware on user devices, and obtaining relevant key information from virtual meetings by bugging personal assistant devices The authors have outlined several challenges in the smart home sector like effectively detecting intrusions for smart homes, defining metrics that are able to classify privacy for smart homes and establishing support for the victims involved in the breach of smart homes.

Kholidy (2021) [6] has outlined the need for CPS to be automated and self-reliant and to have the capability to analyze risks and apply an appropriate response strategy to counter the ongoing attack. The researcher has stated that the current practices in power systems are not appropriate as they neglect cyber-security requirements in the design phase and are oblivious to cyber contingencies. The author has proposed an Autonomous Response Controller (ARC) to retaliate against cyber-attacks in the CPS in an autonomous manner.

Song and Moon (2020) [7] highlight the threats posed by Insiders in the Cyber Manufacturing Systems and the difficulty in detecting and tackling these threats. The researchers have identified that insiders in Cyber Manufacturing Systems can access both physical and cyber domains very easily; moreover, it is challenging to predict the threats posed by the insiders beforehand as there is limited understanding of their motivation and psychology and also there is difficulty in tracing back the attack, as the insiders can very easily conceal their digital footprint. The researchers have proposed a blockchain-enabled Cyber Manufacturing System (CMS) application named FabRec which is able to effectively secure against Man in the Middle attack and UDP DoS attack all the while sustaining the system's flexibility as well as connectivity between physical and digital assets.

Gunduz and Das (2020) [8] emphasize the fact that a smart grid is one of the prominent applications in CPS which is found to be very convenient but also poses serious vulnerabilities which have the potential to lead to disasters in order of national security deficits, lives being lost, widespread economic damage and the disruption of public order and therefore the systems must be designed in such a way that they ensure security. The authors have suggested setting up novel protocols or altering the already existing protocols so that it is tailored to the needs of smart grid applications. The researchers have also proposed exploring new techniques from various fields like data mining and machine learning.

Mantha et al. (2021) [9] have stated that with the advent of digitalization influenced by Industry 4.0 in the architecture, engineering, and construction (AEC) industry, several vulnerabilities have been introduced pointing in the direction of cyber-attacks. The authors have stated that the challenges in the AEC are unique as traditional cybersecurity practices cannot be readily adopted in this industry. The researchers have designed a preliminary model for the AEC industry which has the capability to assist stakeholders who do not hold any prior proficiency in cybersecurity.

Phuyal et al. (2020) [10] have identified several security issues existing in the Smart Manufacturing sector, such as System Integration, Interoperability, Safety in Human–Robot Collaboration, Multilinguism, and the Return of Investment in New Technology.

1.3 Threats and Popular Attacks

1.3.1 Ransomware Attack

As the name suggests, it is a malware which attacks the system accessing confidential files or documents and asks for a ransom amount as payment to recover the stolen files. These attacks are carried out using infected websites and attachments, phishing emails, drive-by-downloads, etc. Ransomware attack was first developed in the late 1980s and the ransom amount was supposed to be sent via snail mail. But nowadays attackers ask for payment to be sent via cryptocurrency or credit card. The first known ransomware was PC Cyborg or AIDS where all the files in the C: directory would be encrypted. But the encryption used was easy enough to be reversed which posed little threat.

In 2004, a stronger ransomware "GpCode" posed a greater threat which used weak RSA encryption to attack personal files for ransom.

In 2007, Winlock instead of encrypting files, locked the users out of their systems. In order to remove them, payment was demanded via paid SMS.

From 2010 onward, there has been a significant increase in the number of ransomware attacks. Some major attacks are Reveton in 2012, Cryptolocker in 2013, KeRanger in 2016 which affected Mac OSes, WannaCry in May 2017, Petya in June 2017, Ryuk in 2018, Sodinokibi in 2019, and many others [11].

The average cost of a data breach is around $3.86 million. As the attackers can claim a large sum of money from the victims thus, we hear of ransomware attacks more frequently. As per the Federal Bureau of Investigation (FBI), in 2017 around 1,783 ransomware incidents were reported that had cost the victims over $2.3 million.

CryptoLocker has been one of the most profitable variants of ransomware. From September to December of 2013, CryptoLocker had infested more than 2.5 lakh systems and generated a revenue of more than $3 million.

There are various ways in which ransomware can infect target systems. The most common method nowadays is Malspam (or Malicious spam) where malware is sent via email messages. The first-ever known mass mailing spams were Melissa mass-mailing virus in 1999 and the famous "ILOVEYOU" mass-mailing worm in 2000 that infected tens of millions of computers worldwide and caused damage worth billions of dollars. From this incident onward email has been seen as the major source of malware delivery and the incidents have been increasing exponentially. The emails have virus-infected attachments such as PDFs or Word Documents and, in some cases, it also contains links to malicious websites. Social Engineering also helps to trick the target users to open the infected documents. The attackers mimic themselves as government officials like from CBI or genuine sources, and scare the users into paying the said ransom amount to unlock their files and documents.

From 2016 onward, there has been a surge in Malvertising (or Malicious advertising) where online advertising is used to distribute the malware. And upon clicking the websites, the target users are directed to malicious websites, thus leading to an attack. Drive-by-download via exploit kit is a common way used in malvertising these days.

The stages of a typical ransomware attack are [11]:

1) **Deployment**: Target system is breached via phishing emails, malicious websites, etc.
2) **Installation**: Once the target machine is infected, installation is conducted by the ransomware and the attacker takes full control.
3) **Destruction**: Using an activated key, the malware locks or encrypts files on the target systems.
4) **Command-and-Control**: Malicious code in the target system is a client and the attacker who is manipulating the system is a server.

There are mainly **two** types of ransomware:

Crypto Ransomware: Valuable files on the target system are encrypted so that the victim can't access them. The Ransom amount needs to be paid to the attackers in order to get their files back.

Locker Ransomware: Files aren't encrypted but the victims are locked out of their systems. Devices are unlocked once the ransom amount is paid.

Some examples of the most frequent ransomware attacks are Locky, WannaCry, Troldesh, Jigsaw, Petya, CryptoLocker, Bad Rabbit, Ryuk, GoldenEye, and GandCrab.

The steps that can be taken **when a ransomware attack occurs** are to:

- Determine which systems are impacted and immediately isolate them so as to prevent further damage.
- Disconnect the devices from the network, and power the devices off so as to prevent further spread of the ransomware infection.
- Begin the process of restoration and recovery so as to get back to business in a more efficient manner.
- Discuss and analyze with the IT team to develop an initial understanding of the impact the attack has caused.
- Inform the stakeholders and internal and external teams about the steps and actions that would be taken to mitigate, respond, and recover from the incident.
- Check for a decryptor so as to decrypt data.
- Don't pay the ransom amount instead report the incident to the police/IT Cell.

The **preventive steps** that can be followed against this attack are:

- Back up your data as this is the strongest defense against ransomware. All data could be restored so that work progress isn't lost. Moreover, backup files must be frequently tested and checked so as to ensure that data are not corrupted and complete.
- Patch and update your system software as this ensures much fewer potential attacks.
- Educate the end-users on these spams and urge them to create strong passwords and use third-party services and devices much more wisely.
- Invest in cybersecurity technology to safeguard system networks. Such software ensures much stronger security for the devices as they use firewalls and IDSs.

1.3.1.1 WannaCry Ransomware Attack

It was on Friday, 12 May, 2017, when the famous WannaCry Ransomware attack [11] started, which, within a day spread to a reach, affecting over 200,000 computers running on the Windows operating system. More than 150 countries were infected by the ransomware, creating a global panic. The majority of the computers affected were running Windows 7 Operating System, while Windows XP and Windows 2003 systems were the most vulnerable. Particularly, those systems which did not have Microsoft's security update from April 2017 were affected. India, along with Russia, Ukraine, and Taiwan were the most affected countries. Cyence, a cyberrisk modeling firm estimated the damages caused by the attack to be worth $4 billion.

As with any ransomware attack, WannaCry ransomware infiltrated the computers of users, encrypted the files on the internal storage, and imposed a ransom of $300 to be paid with bitcoin to decrypt the files.

Renault, a French car-manufacturing company was one of the major companies affected by this attack; they had halted production in several of their factories as a precautionary measure, to curb the spread of the malware, and its partner company Nissan was also affected when its huge factory in Sunderland had to stop the final production shift before the weekend, and its shared factory in Chennai, India, was also affected [12].

National Health Service (NHS), a UK Government-Funded health care and medical service was forced to remain stagnant for several days after it was hit by the WannaCry attack, an attack which cost the UK upward of £92 million, and about 70,000 devices which includes computers, MRI machines, blood-storage refrigerators, and theatre equipment in several hospitals in England and Scotland.

Other high-profile companies which were impacted include LATAM Airlines, FedEx, Telefonica, and Deutsche Bahn.

WannaCry exploits the vulnerability in Microsoft's implementation of SMB protocol and spreads over LAN by sending specially crafted packets which make the computer execute arbitrary code; for instance, if a user received an email with the malicious file as an attachment, and if he were to open it, it would start to spread through the LAN to unpatched systems and start encrypting files all the while appending the WNCRY extension.

Microsoft, however, discovered this vulnerability and released a patch to fix this issue, but owing to the fact that the patch was just a month old, many systems were still vulnerable.

It has been reported that the National Security Agency of the United States of America discovered this vulnerability and developed an exploit for it called EternalBlue, which was in turn stolen by a hacker group named Shadow Brokers, who released it on the Internet via a Medium Post. WannaCry used EternalBlue to infect computers throughout the globe.

In the Deployment phase, WannaCry exploits the MS17-010 vulnerability and injects the binary file named "launcher.dll" with the help of the EternalBlue exploit and using the Doublepulsar backdoor. It also exploits the "srv2.sys", an SMB driver in the kernel module in order to get access to the LAN and send the malicious payload over it.

"launcher.dll" majorly focuses on two functions, namely "ExtractResource" and "CreateProcessMSSECSVC".

Although their inner working is complicated, the idea is that these functions extract resources to create the executable file and make use of the Windows API for this purpose and finally launch "mssecsvc.exe" as a running file.

The "launcher.dll" is injected into the Isass.exe system process and serves as the loader for mssecsvc.exe. Right before the installation phase, two Windows APIs were run, namely "InternetOpenA" and "InternetOpenUrlA" which were used to query a hard-coded domain name of a website which seemed to be gibberish and very long – this was the kill-switch URL. A successful connection to this URL would cause the executable process discussed earlier to stop and quit, but on the hand, if the connection failed, which it often did, the process carried on by dropping "tasksche.exe".

Now the Installation Phase begins, where the main focus is on "mssecsvc.exe" and "tasksche.exe".

"mssecsvc.exe" is responsible for two functions: dropper and infection, whereas "tasksche.exe" takes care of resource loading and encryption environment settling.

There are two ways "mssecsvc.exe" can run.

- **Dropper Phase**: If "mssecsvc.exe" runs without any parameters, execution goes to the dropper phase wherein the sub-functions are called in order to install the "mssecsvc2.0" service and then create "tasksche.exe", then the resource extraction routine completely extracts the resource binary contents and writes it into "tasksche.exe" which is then drop.
- **Infection Phase**: If "mssecsvc.exe" runs with the parameter "-m security", execution is transferred to the infection phase. Now using the attacker machine, the mssecsvc2.0 service will try to connect to port 445 via the SMB protocol and if the connection is successful, the attacker will check if the Doublepulsar backdoor is properly set up by sending crafted packets which have a particular opcode. If the targeted system does not have the Doublepulsar backdoor properly installed, the exploit code continues onward by initializing the EternalBlue attack and as soon as the Doublepulsar backdoor set-up is verified, the payload which contains launcher.dll, kernel shellcode, and userland shellcode will be directly uploaded.
- **Resource Loader**: The first step in this phase is to create a random name for the folder which is to contain the extracted files; this randomly generated name will serve as the unique ID, then the taskche process will check if there is any command argument "i" is present, which represents the installation node of "taskche.exe"; if it is present, then the task proceeds with the installation mode, creates a folder with id, and copies "taskche.exe" into that folder. The resource zip file named "XIA" is extracted from the resource section and then compressed until the resource loading phase is completed, once that is done, the code unzips the resource file using the password "WNcry@2o17", this resource file contains several WannaCry files.

After the Installation Phase, the process moves on to the Destruction Phase, a file named t.wnry was obtained during resource extraction, this file is pre-processed and then, with the help of "tasksche.exe", it is decrypted into a dynamic link library and then passed on as TaskStart and marks the initiation of encryption. WannaCry primarily used RSA and AES algorithms for its encryption purpose; the process begins when RSA generates a root public key, and the corresponding private key is set by the WannaCry creator, and the private key is encrypted with the root public key. As far as the files are concerned, the encryption routine creates random AES-128 encryption keys for different files, and the keys themselves are encrypted [11].

When the file is being encrypted, the AES-128 key is appended to the encrypted file's header followed by the 8-byte magic value "WANACRY!" and the 4-byte length of the AES key.

A British researcher named Marcus Hutchins aka MalwareTech was responsible for stopping WannaCry.

He discovered that right after the deployment phase, the ransomware tried to connect to a non-existing URL and found that this was the killswitch, which could be used to stop WannaCry. He registered that particular domain name and soon the attacks stopped. The attackers, however, tried to sabotage Hutchins' efforts by launching a DDoS attack on the domain name in order to bring it down, but Hutchins was smart enough to protect the site by using cache to take care of the traffic.

1.3.1.2 Maze Ransomware Attack at Bouygues Construction

On Thursday, 30 of January 2020, when Bouygues Group's construction subsidiary (one of the top 10 Global Contractors, with an annual turnover of about $32 billion in the year 2018, with about half in transportation work and half outside France) [13] was paralyzed by a major ransomware attack carried out by Maze ransomware extortionist gang of Russian origin. The entire computer network had been affected, and all of the company's servers were shut down. After the attack, a ransom of 10 million Euros had been requested, and also at least 200 GB of sensitive data were stolen. Meanwhile, shortly after the attack emails sent to Bouygues Construction returned error messages.

This was one of the first high-profile cases of cyber-attacks on major construction firms like Bouygues UK, Bam, and Interserve.

This French organization reported that Bouygues Construction-related servers found in Canada were particularly affected by malware encryption and had a negative impact on the company's operations worldwide. Shortly after the incident Microsoft and McAfee security experts were pressurized to investigate the incident and retrieve information from hackers. In addition, IT staff in Bouygues did not comply with the above-mentioned demands of hackers.

In a statement issued on the 5th of February, the company said it had taken steps to return its information systems to normal use.

Maze, also known as **ChaCha**, was earlier a type of ransomware that was related to online extortion campaigns. If the target organizations denied paying the requested ransom or demands, Maze operators would post their names on a public forum and blackmail them with stolen data as a proof of their successful breach. Maze ransomware attacks commonly try to mimic government organizations and various other governing entities to encrypt and steal sensitive information before asking for the ransom amount from their targets. Typically, Maze is employed as an element of a multifaceted attack strategy. In an extortion campaign, this is rarely the first step. But is usually a part of the second or third step.

While some types of software often run spam campaigns through social engineering or email to gain illegal access to the targeted system. Whereas, Maze ransomware uses exploit kits in drive-by downloads. Exploitation kits usually include a collection of known software vulnerabilities that serve as a utility tool for exploits.

Although there is nothing new about exploit kits or their use, these are mostly used by Maze operators. Maze uses a tool which is dubbed "Fallout" and is a kit which comprises various different exploits identified on GitHub, including the exploit CVE=2018-15982 in Flash Player. The exploit kit Fallout uses PowerShell to complete the dump instead of using the web browser to launch its payload.

After gaining access to the target system, Maze will be encrypting the data, leaving access to its owner. It will then release the encrypted data to threaten it by publicly disclosing it and leaving a digital note behind the victims, which contains steps to make the requested payment before the deadline.

To protect the organization from the attack and exploits of Maze ransomware attacks, there are several steps that can be taken to ensure security. Establishing off-site backups is important so that if the data are locked, the organization can still work to retrieve required data if necessary. All corporate computers must use the latest security solutions and use the latest security features available for any recent risk identified by the security service provider. Protocols such as Multifactor authentication must be put in place and employees should be properly trained in the techniques used by attackers to enter companies' network architecture so they can report suspicious activities if any and take necessary steps to prevent such attacks. Two-Factor Authentication can be used for small-scale uses.

In order to defend from Maze ransomware attacks, the most effective way is to encrypt the sensitive data that disables unauthorized access if any. A secure platform must be made for enterprises that is easy to use for the clients and staff. With powerful end-to-end encryption, data could not be accessed by anyone without the necessary permission, which is a great authentication method.

The following are some of the **mitigation techniques** that can be used to prevent such kinds of attacks:

- User Education
- Backup of critical data
- Continuous security monitoring
- Keeping updated system patched configured in a secured manner
- Restricted administrative access

Construction is one of the few industries that is not driven by data production. Most construction firms don't have large IT departments and have little expertise in managing information security to safeguard from attacks.

However, recent attacks have exposed construction companies to a lot of new threats, making the need for cybersecurity a critical risk management consideration to tackle malicious intrusion. According to some experts, it is predicted that cybercrime would cost businesses around $6 trillion per year on average in 2021.

1.3.1.3 LockerGoga Attack on Norsk Hydro

Norsk Hydro is a Norwegian company based in Oslo, focused on Aluminum and Renewable Energy. It has operations in over 40 countries and is one of the well-known multinational companies in its field.

On March 18, 2019, Norsk Hydro was the victim of a ransomware attack caused by the ransomware LockerGoga that resulted in the loss of over $75 million dollars and the shutting down of production in at least 170 of its plants ranging in the countries of Norway, Qatar, and Brazil; ultimately, this attack affected the entire company with its 35,000 employees across 40 different countries by locking files on hundreds of servers and PCs. The production was forced to be completely offline as 22,000 computers were affected by the ransomware. It took more than a month to return to the normal level of operations.

It all started about three months before the attack when one of the employees opened a malicious email sent by the attackers, which gave them access to the IT infrastructure and they were able to systematically implant LockerGoga, a form of ransomware that encrypted files on servers and devices all over the company.

A ransom note was placed on the screens of the affected computers which stated that the files were encrypted with the strongest military algorithms and only the attackers had the mechanism to decrypt the files, a ransom amount wasn't mentioned but it gave the suggestion that it was increasing by the minute, and as soon as there was contact between the company and the attackers, the ransom amount would be disclosed; however, the mode of payment was to be in bitcoin. After an emergency meeting conducted by the executives of the company, it was decided that they would not be paying the ransom and instead restore the data which was safely stored in their backup server, they then contacted Microsoft's cybersecurity team for support and publically announced about the breach that took place in order to create awareness. Microsoft's Detection and Response Team (DART) deemed the case to be of "maximum severity' and upon conducting a thorough examination, found that an email attachment sent by a trusted customer of the company to an employee was weaponized by the attackers;

this attachment was embedded with a payload that could install a Trojan software on the employee's computer, and soon the virus was compromising regular users of the Norsk computer network before gaining administrative credentials, which allowed the virus to take control of the entire IT infrastructure. The attacker in control pushed the ransomware by using the company network [14].

In the aftermath, DART employees supported Norsk Hydro in restoring their servers whilst improving its security and also conducted awareness and education sessions which provided useful knowledge about the threats that are ever-present and the common behaviors of the attackers, so that such an incident could be avoided in the future and the company's Chief Information Officer pledged to take the learnings from the incident and develop an improved incident response. The company's response to the incident was applauded worldwide and was deemed to be the most ideal response in such a scenario.

As a precautionary measure, the IT department of the company had shut down its servers and network to limit the spread of the virus, everyone was asked to disconnect from the company network and turn off company devices and the entire staff was conducting their day-to-day activities by using pen and paper and plants reverted back to manual procedures in order to continue production [15].

As per the analysis conducted by Nozomi Networks, LockerGoga follows the traditional methods of ransomware attacks wherein it encrypts the files with targeted extensions and then places a text file containing instructions for recovery of the encrypted data. This malware however did not possess the characteristics of self-spreading to multiple targets there was some evidence which indicated that the malware possessed characteristics that would enable it to avoid detection or delete itself from the system, provided the risk of being discovered by sample collections. The malware was seemingly designed to cause disruption to normal services as per the initial examinations. After thorough analysis, it was found that the malware had the capability to run without external support from servers and after its execution, it relocated itself into the %TEMP% folder without obscuring its details. Initially, it was believed that the ransomware only had the ability to encrypt files with extensions like.doc, .ppt, .pdf, .xml, .sql, .js, .py, etc., but it was found that it also possessed the ability to encrypt executable files and.dll files which put the Windows operating system at risk.

The infection vector of this ransomware is usually through phishing or active directory services, the ransomware was found to be written in C++ leveraging popular libraries such as CryptoPP to make use of AES or RSA algorithm simply by re-implementing them. Once the ransomware starts its execution and moves to the %TEMP% file, it runs two commands, one which launches the master, and the other which launches the slave process and uses the predefined file extension list.

The master process is responsible for logging off all active sessions besides the one it is attacking and then hardcodes a password for all the accounts with administrator privileges; it also searches for the files which need to be encrypted and is responsible for disabling the network interface.

The slave process starts encrypting those files that match the targeted extension by using symmetric and asymmetric algorithms such as RSA-4096 and AES-256, and the whole process takes place in offline mode and thus uses embedded public keys of RSA algorithm and once the encryption process is completed, the malware forces a logoff, the administrator password is changed to "HuHuHUHoHo283283@dJD". There is no reported way to decrypt files encrypted by LockerGoga and the only remedy is to restore the files from the backup.

LockerGoga can be detected if there is a file with the extension ".locked".

Additionally, LockerGaga has also affected Altran which is a consulting firm based in France, and Hexion, a US-based chemical manufacturing company.

The best practices that could be employed are keeping a regular backup of files, segmenting networks and data to limit exposure of sensitive data, disabling the use of obsolete components that could be used as an entry point, and securing email gateways.

1.3.2 Social Engineering Attack

The use of the term Social Engineering dates back to as early as the early 20th century when it was used to depict smart methods that would be able to solve social problems. However, with the passing of time, especially after the Second World War, the meaning of this term took a hard swing to the negative side and was generally found to be associated with politicians who used stereotypical designs as an agenda to have an advantage in the elections.

In modern times, the negative meaning of the term still persists, but its usage now rests heavily with cybersecurity wherein it is known as a manipulation technique to accomplish a broad range of malicious activities by exploiting human error [16].

Social Engineering is one of the rare attacks that is considered a non-technical attack, but it has the capability to join forces with other technical attacks, which makes it a formidable foe.

Recent statistics show that around 98% of cyber-attacks have reliance on social engineering and over $5 billion has been lost in the past few years raising serious concerns.

What makes this attack very dangerous is not its very complicated nature but on the contrary, its subtle and relatively simple nature wherein it targets the weakest factor in any company or industry, i.e., humans; people today are more available and more connected than ever before, making them highly susceptible.

The **two main categories** of social engineering are:

1. **Technology-Based Deception**: In this approach, the user is tricked into perceiving that he is interacting with his intended computer system, which leads the user to readily give out classified knowledge.
2. **Human-Based Deception**: Here, the cunning and crafty nature of the attacker comes into play, using which he exploits the victim and manipulates him [17].

Steps Involved in Social Engineering

1. **Gathering Information**: With the emergence of social media, this step has become a relatively easy task, the attacker can easily do a "background check", and in a short amount of time, come up with a general profile of the intended victim, by analyzing their posts, pictures, interests, experience and the plethora of information available on their social networking handles.
2. **Fishing** for trust with the user: The attacker approaches the soon to become victim, and using manipulation techniques and the information discovered in the previous stage, is able to win him over.
3. **Exploitation**: With the help of the previous two steps, the attacker has now befriended the victim and can finally manipulate him by taking adv.antage of the emotional state of the victim and getting the desired information he needs.
4. **Exit**: Now that the attacker has exploited the user, he disappears as mysteriously as he appeared [18].

Different types of Social Engineering Attacks

1. **Pretexting**: In this attack, the attackers try to impersonate a trustworthy or an authority figure and create a fictitious scenario, and obtain needed information
2. **Quid Pro Quo**: Once again, the attacker impersonates himself as either a technical expert who extracts sensitive information from the victim or by calling an actual technical expert and pretending that he is facing technical issues, and needs sensitive information.
3. **Phishing**: It is an attack that acquires personal data and other sensitive information by impersonating as a credible source via various modes of digital communication, most prominent of which is electronic mail.
4. **Baiting**: In this attack, the victim receives an email or advertisement about promised rewards in terms of valuable items or cash prize, and is prompted to enter his sensitive information, which is posed as vital.

The users receive a link, and upon clicking are taken to a look-alike website which is convincing enough for the users to readily hand in their sensitive information. Other types of Phishing include Spear Phishing, Vishing Whaling Phishing, etc. [17].

1.3.2.1 How a Possible Social Engineering Attack Occurs

A technical attack can very easily be conducted by using the Social Engineering Toolkit that is present by default in Kali Linux.

Kali Linux is a Debian-based Linux operating system maintained by Offensive Security and is designed for the use of network analysts and penetration testing. It contains a plethora of tools for the purpose of ethical hacking and the Setoolkit a.k.a Social Engineering Toolkit is one of them.

It is easily accessible by simply typing "setoolkit" in the terminal, and is also present in the apps menu.

Once opened, the user is presented with six different options and a seventh one for exiting the software, upon selecting the first option, namely Social-Engineering Attacks, we are presented with eleven different attacks. In this example, we concentrate on a simple Credential Harvesting Attack. We select options Website Attack Vectors ≥ Credential Harvester Attack method ≥ Site Cloner.

Now all that's left is to provide the IP address of the machine that is performing the attack and the URL of the website that needs to be cloned, a fraudulent link is generated and sent over to the user, who can easily be tricked by the uncanny resemblance, and enter their credentials, once this is done, the credentials are then redirected to the attacker's machine server and he is able to obtain the required information with little effort.

1.3.2.2 Preventive Steps

The industries should concentrate on raising awareness of social engineering attacks, as it is mostly the exploitation of human nature. There should also be restrictions on the amount of information any particular individual has access to. Proper email-filtering techniques should be employed and users should never click on suspicious links or emails that ask for personal details. Only official sites which have https protocol enables should be used. Strong passwords are a must and along with it, if a two-factor authentication is used, it further increases the security. An enhanced blacklist should be set up so that employees are not able to access phishing websites, and as far as Trojan-based attacks are concerned, usage of USBs should be disabled.

1.3.2.3 FAAC Whaling Attack

Fischer Advanced Composite Components AG (FAAC) is an Austria-based aerospace manufacturing company whose customers include Airbus, Boeing, Rolls-Royce, Siemens SAS, and Mitsubishi Heavy Industries among others.

In 2016, FAAC was hit by a whaling attack; this scam is also known as "CEO Fraud Attack" or "Fake President Attack" which resulted in the loss of over €50 Million; the financial results however showed that the company soon realized that this ordeal was a scam and managed to stop the wiring of €10.9 Million but as far as the remaining money was concerned, it was long gone to Slovakia and Asia.

In almost every whaling attack, the common pattern is that a fraudster sends an email to an employee in the finance department of the company by posing as the CEO or any other senior board member of the company intimating a money transfer out of the company. In this case, the attacker sent a fake email to an employee in the finance department in which he impersonated the then CEO Walter Stephan. The email asked the employee to transfer €50 Million for the supposed acquisition project of the company and there was a sense of urgency in that email. The money was subsequently transferred to a foreign account.

This was a huge loss to the company as it managed to engulf the entirety of the previous year's profit and then some which put the company at a significant deficit.

FAAC's supervisory board, after a 14-hour long meeting decided to fire Walter Stephan, who was the CEO at the time and had been for the past 17 years; the board concluded that Mr. Stephan had severely violated his duties with respect to the "fake president incident". The CFO (Minfen Gu) was also fired and both Walter Stephan and Minfen Gu were sued by FAAC for $11 million in damages [19].

Investigators arrested a man in Hong Kong in August in response to the FAAC Whaling Attack; however, neither the location details of the accounts nor details of the arrests were released.

The FBI, in the same year, warned about the steady increase in such attacks which has resulted in the loss of over $2 billion and a Belgian bank lost $75 million around the same time to a similar scheme. There is no indication that this type of attack is going away anytime soon; on the contrary, it's only increasing, and, therefore, there is a need to establish severe protocols which can prevent such a mishap in the future.

1.3.2.4 The 2005 DuPont Insider Attack

DuPont is an American company in the chemicals industry, the company is listed in the Fortune 500 and was at one point, the world's largest chemical company in terms of sales.

In 2005, DuPont was setback by a massive insider attack which resulted in the loss of $400 million in trade secrets, and was carried out by Yonggang "Gary" Min, and as soon as DuPont discovered that a huge amount of confidential and proprietary data had been stolen by Min, it informed the FBI.

Gary Min worked as a research chemist at the Delaware-based company from 1995 to 2005 before joining a competing company based in Asia named Victrex in December 2005. In the months leading to his departure from DuPont, he would become the single most active user of the company database, downloading 15 times the data as compared to any other employee, and is believed to have viewed over 16,000 sensitive documents and downloaded an additional 20,000 of the sensitive information present in DuPont's e-library. Min's acts were discovered only after he signed in his resignation to the company, and the discovery is believed to be either a cause of analyzing database logs or by the automatic alert system of the database. The amount of documents taken was so large that Min had to separately rent out a storage facility and an apartment just for the purpose of storing the stolen documents. After starting work at Victrex, Min transferred 180 of the documents he had procured from DuPont into the company laptop given by Victrex [20]. In a statement, Min had revealed his intention of stealing the trade secrets of DuPont was motivated by the idea of handing it over to his new employer, Victrex, which was a competing organization. Victrex, however, has denied even seeing the documents and went so far as to assist police officials in the arrest of Min.

In 2007, Min was sentenced to 18 months in prison and a $30,000 fine; he faced a light sentence as the prosecutors could not prove what Min did with the compromised documents, and it could not be proved that the stolen information was shared among competitors.

Min's actions were a violation of integrity and trust placed in him by the higher-ups of the company and showed that it was relatively easy for an insider with required privileges to compromise sensitive data. Furthermore, there have been several other cases of insider attacks on DuPont; in 2009, Hong Meng stole sensitive proprietary information related to OLED technology and planned to take it to his university of graduation in China, and the company found out about his actions only when it reviewed the contents of his hard drive before his departure to China. Another recent case was of Michael Mitchell who had proprietary data of DuPont in his home computer, and after he joined a Korean competing company, he leaked this sensitive information to them, resulting in a huge loss for DuPont [21]. The severe actions taken against the criminals and awareness created by DuPont go a long way to show the seriousness of the otherwise neglected matter of insider threats.

These incidents go to show that it is not only the threats that arise from the outside that are dangerous but the insider threats are just as dangerous, if not more in certain scenarios and the only way for a company to avoid these unwarranted losses is to ensure security for these threats by frequently analyzing the database logs and setting appropriate thresholds based on the employee's requirements and other proactive measures need to be taken seriously in lieu of this underlying threat.

1.3.2.5 Toyota Boshoku BEC Attack of 2019

In September of 2019, a European subsidiary of Toyota Boshoku Corporation which itself is a subsidiary of the automotive titan Toyota Motor Corporation was hit by a Business Email Compromise attack resulting in the loss of around $37 million.

The hacker cleverly disguised as a business partner of the company, sent a seemingly convincing email which at first impression might have been indistinguishable from the real thing, asking one of the employees in the finance department to wire $37 million to the bank account which was in the control of the hacker [22]. The hacker had undoubtedly excellent social engineering skills, given the fact that they were able to convince an intelligent employee of the company to wire out a large sum of money and is speculated to be a result of a cleverly named email address which was in close agreement with the actual business partner's email barring the few typos which would have existed but gone unnoticed. There was also speculation that the hackers had a well-painted canvas of their target owing to the fact that Toyota had been a victim of two cyber-attacks earlier in the year, and it is suspected that the hackers could have obtained the required credentials of the important members of the corporation but this speculation does not rule out the chances of spoofing which has historically had quite an impact.

The attack was forecasted to setback the earnings report of the Toyota Corporation for the first quarter in March 2020 which would inevitably lead to a negative impact on the stock prices of the company [23].

The FBI reports that over $5 billion have been lost due to BEC attacks in the last few years and it has been found that 75% of companies are exposed to BEC attacks on an annual basis.

Generally, the hackers have a polished approach for targeting large companies wherein the plan is set in motion months or even years in advance, and it begins with a thorough reconnaissance and knowledge formation of the target, and for this task to be successful, the hackers often perform phishing attacks to plant malware on the employee's system and are thus able to sniff the email conversation and understand the pattern of communication that is carried out and pounce on the target when the right opportunity presents itself and with their expert social engineering skills, they are able to come out on top and a lot wealthier. Despite having strict cybersecurity practices, the attackers were able to take an alternative route and exploit the highly vulnerable human factor, i.e., take advantage of the employees of the organization.

For a company of Toyota's stature, the transfer of an amount like $37 million was nothing stupendous and therefore went undetected until it was too late. Since they had effective security measures in place ensuring that any external attack would be difficult, the internal factor was neglected and served as a good opportunity for the attackers and also serves as a wake-up call for other organizations in the CPS.

The preventive measures that can be taken against BEC attacks are educating the employees about BEC attacks and studying their recurring patterns in order to be more cautious, implementing a multistep approach for transferring large funds which would require the authorization of multiple supervisors, establishing email security at the mail server, and protecting it from phishing attacks. Many companies have begun implementing two-factor authentication and have established separate lines solely for the purpose of money transfer, it has also been observed that finance departments are now being closely aided by the security department and early fraud detection technology is being implemented [23].

1.3.3 DDoS Attack

DDoS attack [24,25] disrupts the normal traffic of a targeted server, service, or network by attempting a malicious attack to surround the network infrastructure with a sudden flood of Internet traffic. Basically, a DDoS attack creates an unexpected traffic in the network. The load on the server varies from 620 Gbps to 1.1 terabits per second. The average loss incurred by this attack was USD 417,000.

Effectiveness is achieved by using a lot of compromised computer systems as sources of attack traffic. The machines that are exploited are computers and various other network-dependent systems such as IoT devices. These individual devices (bots) create a network known as botnet. The attacker directs attacks by sending instructions to each bot once the botnet has been established. DDoS is one of the most profitable attacks and yet affordable from the attacker's point of view.

The first DoS attack took place in 1974. Till now, DDoS is one of the most lethal attacks. From 2007 onward, there has been a sudden rise in the number of DDoS attacks. Moreover, the emergence of IoT industry was the turning point for DDoS attackers. The amount of impact it made on the industry is humongous, the face that it was harnessed in hybrid attacks which makes it even more dangerous. It has been already more than 25 years and yet, DDoS still remains one of the favorite choices for the attackers.

A total of six banks, which included JPMorgan Chase, Bank of America, US Bank, Wells Fargo, Citigroup, and PNC Bank, were attacked on 12 March 2012, which, to date, remains one of the most impactful attacks on a sector which is finance in this case. This attack created a sudden 60 gigabits of DDoS attack traffic per second on the network of each of the above six companies.

Mirai Botnet also collapsed the entire Internet connectivity of the East Coast of United States. The self-multiplying tendency of this attack creates a large-scale impact on the entire network. Similarly, the GitHub attack on 28 February, 2018, which lasted for 20 minutes, clocked a DDoS traffic of 1.35 terabits per second.

The most recent attack has been the Amazon Web Services (AWS) attack in February 2020 [26]. This lethal attack amplified the victim's IP addresses by approximately 56–70 times. This attack which lasted for three days was struck with a massive traffic of 2.3 terabytes per second.

As per the prediction given by Cisco, the total number of DDoS attacks would nearly double from the 7.9 million witnessed in 2018 to somewhere around 15 million by 2023.

The reason why attackers choose a DDoS attack is because the scale of impact that this attack causes is unmatched. As per some surveys, the investment is considerably negligible to the amount of profit that the attackers make out of this powerful attack. The primary goal of it is not to steal information but to slow down or take down a website by a sudden surge in the network traffic. Moreover, website visitors' information could also not be stolen by these DDoS attacks.

Once the website resources are overloaded, the attackers find a way of extortion and blackmailing. However, it may have some other motivations as well, which include political, terrorist, hacktivist, and business competition. Any competitor with a financial or ideological motive could easily damage an organization by launching a DDoS attack against it. Such contract assignments are taken over by the DDoS attackers. The amplification factor also plays a huge part as around 1,000 bots are used to take down a single server.

First of all, the hackers identify their targets and analyze the information and features regarding the network architecture and other necessary things. The DDoS attack would test the limits of a web server, network, and application resources by directing sudden surges of fake traffic. Some attacks are short bursts of malicious requests that are done on vulnerable endpoints such as search functions. An army of zombie devices or bots is used to form a network called a botnet. These botnets mostly consist of various compromised systems such as IoT devices, websites, and computers.

The botnet would attack the target and deplete all the application resources once a DDoS attack is launched or initiated. Effective DDoS attacks can prevent users from accessing the website or slow down resulting in financial losses and operational problems that are being intended by the attackers and the rival companies in most cases.

This attack completely reduces server resources and increases website load time. When a website is hit by a DDoS attack, it can suffer from operating problems or hit the server completely by overloading server resources such as memory, CPU, or the entire network. Most DDoS attacks from botnets are taught to hackers of many vulnerable IoT devices, including

Internet-connected IoT devices such as security cameras, smart televisions, home lighting systems, appliances, and smart refrigerators.

The apparent increase in DDoS attacks is mainly due to the complete lack of IoT device control, which makes them very good bottlenecks. A hijacked group of IoT devices with unique IP addresses can be redirected by attackers to make large and vicious sudden attacks targeting websites, creating DDoS attacks.

There are usually **three types of DDoS attacks**:

1. **Volume-Based Attack**
   A huge number of requests are sent to the target system. The motive is to create a sudden flood of network capacity. Various ports of the system are used to send requests. Once the system is compromised, the server's IP addresses are spoofed. This type of attack involves numerous systems which further amplify the impact of this attack.
   Examples include UDP floods, ICMP floods, Ping floods, etc.
2. **Application-Based Attack**
   Attackers exploit the vulnerabilities in the application software which leads to the crash or hanging of webserver. An application-based attack sends partial requests to the server so as to make the entire database connection busy such that all legitimate requests are blocked to the server. It is one of the most serious and sophisticated types of attacks.
   Examples include HTTP flooding, BGP Hijacking, etc.
3. **Protocol-Based Attack**
   In this attack, the motive of the hackers is to exploit the server resources or load balancers so as to communicate with each other. It is likely that the servers are being made to wait for some non-existent responses during some handshaking protocols by the malicious packets created by the attackers.
   Examples include Ping of Death, SYN Flood, etc.

1.3.3.1 Prevention Steps

The following are some of the important prevention steps to safeguard from DDoS attacks:

- **Develop a DoS Response Plan**: The organization must give high priority to designing a Response Plan. The key elements of this include
  - Creating a checklist of systems
  - Formation of a response team
  - Notification and other escalation steps need to be defined
  - Internal and external contacts list must be included

- **Secure Your Network Infrastructure**: A strong multi-level strategy needs to be designed for network infrastructure protection. It needs to include some advanced Intrusion Prevention Systems which combine with VPN, load balancing, Firewalls, etc. All the software and hardware requirements need to be patched and updated on time.
- **Practice Basic Network Security**: Basic protocols need to be followed by the organization, such as maintaining a strong firewall and performing other methods such as anti-phishing and maintaining strong login credentials.
- **Maintain Strong Network Architecture**: The network architecture must be competent enough to handle large traffic and also be able to stabilize itself in case of failure of network resources.
- **Leverage the Cloud**: In order to stop the attack, cloud-based services need to be used to analyze the magnitude of the attack so the design a tactic to safeguard the network system. Hybrid environments shall also be designed to achieve the right balance between security and flexibility.
- **Understand the Warning Signs**: The organization needs to be vigilant and observe the signs of a probable DDoS attack so as to timely perform the designed action plan.
- **Consider DDoS as a Service**: It provides flexibility to combine existing and third-party resources such as cloud and dedicated server hosting to ensure that high-security standards and compliance requirements meet all aspects of security infrastructure.

If a DDoS attack happens, the following needs to be done:

- Alert the main shareholders
- Security Provider needs to be intimated
- Defense measures must be activated
- Monitoring the progress made by the attacks
- Defense Performance is to be assessed

1.3.3.2 MIRAI Botnet Attack

On 12 October, 2016, a huge DDoS attack left a large part of the Internet unreachable on the East Coast of the United States. The attack, which officials believe was a conspiracy by a rival nation, was in fact the work of the Mirai botnet. On further investigation, it was revealed that the attack revealed 49,657 unique IPs which were used to host the Mirai-infected devices. Most of these devices were CCTV cameras which are the most popular choice of the DDoS botnet attackers. Other targeted devices were routers, DVRs, etc. All these devices were poorly secured. This attack was successful because most of the Smart IOT device users were relatively inexperienced and thus was able to breach 600,000 IoT devices. It scans the major Internet blocks to detect open Telnet ports and tries to sign in to devices using default passwords. That way the attackers were able to create a botnet army. This attack, initially had a few major ambitions but later became much stronger than previously thought.

Paras Jha and Josiah White founded Protraf Solutions, a company that provided services to reduce DDoS attacks. They were basically running a racket where their company offered DDoS mitigation services to the organizations where their malware attacked just a case of insider theft.

Mirai is malware that infects smart devices running on ARC processors, converting them into a network of remote-controlled bots or "zombies". This network of bots is collectively called a botnet. It is used to launch DDoS attacks on targets and devices to harm them.

Mirrors is a self-expanding worm, with a dangerous system that repeatedly replicates itself by detecting, attacking, and infecting vulnerable IoT devices. It can also be thought of as a botnet because infected devices are controlled by a network of command-and-control servers (C&C). These servers instruct the infected devices on the sites that are to be breached in a sequential manner. Once the Mirai is able to discover open Telnet ports, it tries to infect the unprotected devices by brute-forcing the login credentials [26].

Mirai consists of **two key** components:

- Replication Module
- Attack Module

1.3.3.2.1 Replication Module

The replication module increases botnet size by ensuing vulnerable IoT devices. It scans the entire Internet for potential triggers. Once a vulnerable device is compromised, the module informs the servers so as to infect the target device with the latest Mirai payload.

Initially, Mirai relied solely on a modified set of 64 authentication credentials such as login/password combinations that are commonly used by IoT devices to compromise devices. The attack was not technologically advanced but proved to be very effective and deterred more than six lakh devices.

1.3.3.2.2 Attack Module

This module is responsible for carrying out DDoS attacks against the targets specified by the servers. It uses most of the code using DDoS techniques such as HTTP flooding, UDP flooding, and various other TCP flooding methods. This wide range of methods enabled Mirai to carry out application-layer attacks, volumetric attacks, and other TCP state-exhaustion attacks.

1.3.3.2.3 Prevention Methods

The capabilities of the devices need to be limited. We need to check whether Remote Access is required or not. If Remote Access is needed then a Strong Device Authentication must be implemented. It is to be ensured for Administrative Users and Services.

Ensure only Authorized Software and Firmware Updates. Like other botnets, Mirai also uses well-known methods to attack and compromise devices. It also tries to use the login credentials that are known to operate on the device and take it out of all sensitive information.

Botnets attack the systems that haven't been updated with current security patches. The exploits keep on targeting the business hardware on a regular basis. Thus regular security and hardware update is a must.

We must change our login details (like username/ passwords) quickly. Mobile applications or web interfaces can be used to change the credentials for the routers. And for other devices, one needs to sign in with their default credentials and consult the device manual if required for assistance. All login/password combinations are targeted by the new Mirai variants.

If the latest firmware isn't supported, then the user needs to change the device and install a newer device with all newer hardware and software setups

Replacing the devices seems to be drastic, but to ensure zero vulnerability and high security, replacement is the best option available. Botnets like Mirai can't be destroyed as they would be around us and would become stronger and tougher to defend. Thus, in order to protect the devices, all steps need to be followed and ensured and, if we can successfully protect our own devices, then the rest of the Internet can also be safeguarded.

1.3.3.3 2020 AWS Reflection Attack

Amazon Web Services was hit by the largest publicly reported DDoS attack in February 2020; the DDoS attack was capped at 2.3 Tbps and it was larger than any previous attack by a substantial margin. Amazon was able to successfully attenuate the attack with the help of AWS Shield, which is a protection service offered by AWS to prevent DDoS attacks [26].

The attack was based on a Connection-less Lightweight Directory Access Protocol which is also known as CLDAP DDoS reflection attack, this attack was supplemented by amplification attacks. The amplification achieved can be of the order of 100 times the normal traffic capabilities generated by the original attacks and also serve the purpose of hiding or spoofing the attacker's trail. The attacker exploited the vulnerabilities in the CDLAP server and heavily amplified the amount of data that was normally being sent with the motive of disrupting operations which resulted in three days of elevated status of the threat and as reported in the AWS Shield threat report of the first quarter of 2020, this magnitude of attack was unforeseen and crippled any previous network attack that AWS has suffered, besting it by a significant 43%; however, AWS came out unscathed as they had proper security mechanisms in place, which were strong enough to even suppress such a powerful attack, and some credit has to be given to the approach taken by the AWS team after they had been hit by a DDoS attack in the previous year, which resulted in AWS experiencing problems for eight hours.

CLDAP is based on the User Datagram Protocol's directory lookup protocol and in the design phase was originally supposed to lower the connection overheads but the design also presented some underlying vulnerabilities such as anonymous access, lack of confidentiality protection, and missing integrity protection, and one of the key drawbacks present was that whenever there was a misconfiguration of CLDAP server, it responded to even spoofed requests and it also had a very high amplification factor of 56–70 which is sought out and exploited by attackers.

The nature of the attack is straightforward – it takes advantage of the connectionless nature of the User Datagram Protocol and sends spoofed requests by spoofing the soon to be victim's IP address to the identified open servers on the Internet which are misconfigured, the server in response floods the victim with a large amount of amplified response and the victim is left helpless as their normal services are eventually denied owing to the overloading of the server with an abundance of requests and given the amplification factors of these servers, the attacker can get a lot done just by sending tailored short queries which result in a massive response from the server, and given the distributed nature of the attack, it goes without saying that there are several botnets spread out, spoofing the victim's IP address, which continually repeat that query. This act renders the whereabouts of the attacker impossible to locate, meanwhile the response would not be as simple as blocking a particular connection as multiple botnets are leveraged for this attack and since the attacker's IP address is spoofed to be the same as the victim's IP address, the server is unable to distinguish between them, and hence the name Reflection attack is suitable.

The attack occurred on UDP port 389, and a good countermeasure to avoid such a DDoS attack would be to set a traffic limit on the particular port and adapt encrypted LDAP configurations, no open server should remain misconfigured as it invites future DDoS attacks, Regular Expression filters should be in place which can detect the common pattern of these attacks and prevent them at an early stage.

1.3.3.4 2018 GitHub DDoS Attack

On 28 February 2018, one of the largest DDoS attacks took place and it attacked GitHub. The attack lasted for 10 minutes. In the first portion of the attack, a traffic of 1.35Tbps struck between 17: 21 UST and 17:26 UTC, and in the second portion, from 17:26 to 17:30 UTC with a spike of 400Gbps.

At the time of its peak, 126.9 million packets per second was the traffic and this fake request led to the collapse of GitHub(dot)com. UDP-based Memcached traffic was the reason behind the amplification of this attack. It is a database caching system which is designed to accelerate websites and networks. It floods the targeted system with a sudden influx of traffic to crash or hang their servers [27].

Akamai Prolexic was called in to stabilize the situation. Upon investigation, it was found that Ip Addresses were spoofed. The traffic was rerouted and the malicious patches of code were removed and the blocked data were recovered. Thus, after eight minutes, the assault was stabilized and the DDoS stopped after tackling all kinds of anomalies. The Chinese government had been widely suspected to be the guiding force behind this attack. As there were a few projects which the Chinese government had repeatedly asked to be removed from the website. Due to unsuccessful censorship, the intent was to attack and pressurize GitHub into eliminating those projects [28].

Steps taken to prevent similar future incidents include the following:

1. Various automation steps had been taken while keeping in mind the future conditions. Less human dependence would ensure smooth and hassle-free use of this novel open source.
2. A monitoring infrastructure had been proposed that would enable the DDoS mitigation providers to measure and analyze the impact in times of attacks so as to safeguard themselves. This would ensure a reduction in Mean time to Recovery (MTTR) [29].
3. The network architecture had been improved and expanded with Multiple new layers with added features to identify and mitigate new attempts of attacks before the workflow of the website is affected. Regular backups of the data need to be done to ensure safety and security. In order to uphold the trust of its clients and users, the documents and projects which have been stored in their website includes a better and effective security layer.
4. Moreover, regular updates in software enhancement to build better detection systems and enhance streamline response have been promised.

1.3.4 Man in the Middle Attack

Man in the Middle attack is an attack where the communications between two parties are intercepted by an attacker with a motive to steal personal information which includes login credentials, bank account details and credit card numbers, etc. Spying on the victim in order to sabotage communications or corrupt sensitive data might also be their intention. Most of these attacks are targeted at the financial sector, SaaS businesses, E-Commerce sites, and some other web applications which involve logging and payment transactions [30,31].

As reported by Netcraft in 2016, more than 95% of HTTPS servers are vulnerable to MitM attacks. These attacks are not as common as Ransomware attacks but they are a threat to the organizations. IBM's X-Force Threat Intelligence Index 2018 report stated that around 35% of exploitations directly and indirectly involved MitM attacks. Only 10% of the companies have safeguarded their servers from MitM attacks. The damage caused varies, depending on the attacker's intention and capabilities [32].

It is one of the oldest forms of cyber-attacks which is still prevalent. The oldest known Man in the Middle attack dates back to 1586 when a coup to assassinate the then Queen Elizabeth I was failed by the royal secret servicemen, and the culprits were caught red-handed [33].

After Guglielmo Marconi invented Radio Transmission communication, this primary mode of communication was under a serious threat of MitM attacks. Since then, there has been a surge in the number of incidents [34].

British Intelligence was able to decode the Enigma Code using this attack and was thus able to stop the Nazi threats and attacks. This was the primary reason that led to Adolf Hitler's defeat in the Second World War.

In 2013, Edward Snowden disclosed the NSA's attempts to leak user information using Quantum/FoxAcid Man in the Middle attack system. Lenovo installed Superfish, an SSL Hijacking Adware on their Windows PCs in 2014. Email hijacking/eavesdropping attack on the British Lupton couple in 2015 cost them over 340,000 Euros [35,36]. Such incidents regularly happen where the scale of impact varies.

The reason why this attack is used is that it enables the attacker to easily fool its target and fetch all the sensitive information. Of all the methods available to attack are many which increase the chances of a successful breach. This attack has been used for centuries and has been one of the favorites of hackers since then. The ability to mimic themselves as original servers and service providers is the main advantage which helps them to carry out their attacks without being suspected at all. Moreover, to date, this attack has impacted a vast audience of targets.

The attacker identifies the target and then tries to intercept the connection. In some cases, the connection is secured through encryption. Based on the target and the goal a technique is chosen from a broad range of methods. The attacker monitors the conversation of the target. Spoofing, Hijacking, Eavesdropping, and Intercepting are some of the options available for the attackers. While the conversation between the target and the other end takes place, the attacker intercepts the messages between the peers. Upon successful interception, the attacker steals sensitive info, alters the message, etc. Sadly, the attackers are also able to bypass Multifactor authentication in some cases which enables them to steal credentials.

1.3.4.1 Man in the Middle Attack Types

The following are the various types of MitM attacks:

1. IP Spoofing
   Every device has a unique IP Address. By spoofing one's IP address, the attacker makes the victim think that they are interacting with a website which isn't the case. Thus, it enables the hacker to get all the user information of the victim.
2. DNS Spoofing
   Domain Name Server spoofing is a method that forces the user to use a fake website instead of the original. This allows them to divert traffic from the actual site in order to hold user login credentials.
3. HTTPS Spoofing
   Every URL in the Internet starts with HTTP. But we must ensure that we always use secure websites which are usually identified with the website starting with HTTPS where S means secured. The attacker usually fools the victims by redirecting their browser to an unsecured website. This enables them to monitor the user's interactions with that webpage so that they could steal the personal information which they are sharing.
4. SSL Hijacking
   Secure Sockets Layer (SSL) is a protocol that establishes encrypted links between a web server and a browser. This ensures that the data shared with the server is secure.
   In SSL hijacking, the attacker successfully captures all data passing between the target computer and the server through another computer and secure server.
5. Email Hijacking
   Email accounts of banks and various other financial institutions are targeted so that attackers can gain access to and monitor transactions between the organization and its customers. The bank's email address has shovels to send its instructions to customers. Therefore, the attackers impersonate the bank and assured customers to follow their orders rather than the bank. Hence, the customers are fooled to deposit their money to the fraudsters.
6. Wi-Fi Eavesdropping
   Wi-Fi connections are set by attackers with real audible names. When a user connects to the attacker's Wi-Fi, the user's online activity can be monitored and it is possible to obtain login credentials, credit card details, and other sensitive information. To avoid such dangers, we should not use public Wi-Fi.

7. Stealing Browser Cookies
   A browser cookie is a small piece of information stored by a website on a personal computer.
   Because cookies store data from personal browsers, attackers can gain easy access to passwords, addresses, and other sensitive information. Therefore, attackers often steal these browser cookies.
   Some of the examples of Man in the Middle attacks are Babington Plot in 1586, Cracking the Enigma Code by British Intelligence in the Second World War, Edward Snowden exposing NSA in 2013, Lenovo using Superfish in 2014, and Lupton Couple incident in 2015.

The **prevention steps** that can be followed are:

- Disable your devices to connect automatically to Wi-Fi networks. Make sure you only use trusted Wi-Fi networks.
- The access points need to be encrypted and secured.
- Use VPN to secure your traffic while connecting to an unknown or public Wi-Fi network.
- Avoid sharing sensitive information with an unsecured URL (secured ones begin with https://).
- Always enable Two-Factor Authentication to your accounts.
- Regularly update the OS so that attackers don't exploit the weaknesses of an outdated system OS.
- Installing a trusted antivirus software is a must, followed by regular weekly scans.
- Beware of phishing attempts in your emails and never log in to a link of a mail if it seems suspicious.
- Wi-Fi network should be secured and always use strong and unique login credentials instead of the default ones.

1.3.4.2 2015 Lupton Couple Incident

In February 2015, Paul and Ann Lupton a couple from Britain sold their flat which they had earlier bought for their daughter [35]. After all the negotiations, they received an email from their Solicitor on February 27 which was two days before the completion date, requesting their bank account details so as to transfer the amount of £340,000 into their account.

This entire transaction and fraud took place in Barclays Bank Account. As a reply to Solicitor's mail, Mr. Lupton sent his Sort Code and bank's Account number. The mail was intercepted in the middle by the hackers. They used a sophisticated tool that examines numerous emails to identify the ones which shall contain valuable information. The attackers posed as Mr. Lupton and sent an email to the Solicitor to omit the previously sent account and instead pay in their (fraud) account. This was an email eavesdropping/email hijacking MitM attack. Thus, the solicitor sent the balance amount after deducting his consultation charges. The fraudulent account received £333,000.

After a few days, when the Lupton Family enquired their solicitor about their payment which was due, they came to know about the fraud. Immediately, the fraud was brought under the notice of both Barclays Bank and the police. This event highlighted the lack of resources that firms have to safeguard their client's money. The sensitive and confidential information are under threat in emails as well. On hearing about this incident, various Government security service firms stated that this case was a wake-up call of how sophisticated the hackers have become these days. The Solicitors Regulation Authority (SRA) had made a verdict and warned that financial firms are responsible for protecting their customers' finances and should be compensated if their funds are misappropriated or withdrawn from the customer's account. Although they can't force the firms to compensate the victims.

After this event, Barclays froze Mr. Lupton's account and initially returned £271,000 to them, although they tried to defend themselves and blamed the sophistication of the fraud. And after eight months, their PI insurer (Perry Hay & Co) paid the balance amount and reimbursed the victim (The Luptons).

Although detecting an eavesdropping attack is challenging, the following steps help to reduce the risk significantly [37]:

- **Encryption**: Emails, networks, and communications, and data at rest, in use, and in motion need to be encrypted. Attackers would need an encryption key to decrypt the files. Using a 256-bit which is also known as military-grade encryption is ideal.
- **Authentication**: In order to prevent IP spoofing, we need to authenticate the incoming packets. Using standards and protocols that provide authentication is also suggested.
- **Network Monitoring**: Deploying an intrusion prevention system is a must so as to constantly monitor traffic and identify abnormal activity.
- **Cybersecurity Awareness**: Basic knowledge of cybersecurity needs to be taught to the users. This helps them to tackle the situation better. Using strong passwords in systems and Wi-Fi networks and avoiding public networks are some known facts. Encourage the use of VPNs and Firewalls.
- **Network Segmentation**: Segmenting the network ensures that the data can't be accessed and reached by the hacker. The separate segment partition ensures that traffic can't flow from one segment to the other.

1.3.4.3 2014 Samsung Smart Fridge Man in the Middle Exposure

In April of 2014, Samsung took a second shot at making a smart fridge named Samsung RF28HMELBSR/AA Refrigerator which was priced at $3599 with an 8-inch LCD screen, inbuilt Wi-Fi, and features like phone mirroring, "Kitchen TV", wherein users would be able to connect the Fridge to the antenna's signals and enjoy the broadcast from their kitchen, as well as "kitchen calling" wherein users would be able to make calls directly from the smart fridge. The fridge also has the capability to display the calendar and also the news. However, with the plethora of features that the fridge brought, it also brought with it some vulnerabilities which when exploited could potentially be dangerous.

At the IoT hacking challenge conducted at the DEF CON hacking conference, researchers of the company Pen Test Partners discovered an existing Man in the Middle vulnerability in the RF28HMELBSR model of Samsung Smart Fridge. While the fridge implemented the SSL, its shortcomings lay in the failure to actually validate the SSL certificate, which proverbially opened doors for Man in the Middle attacks [38].

The fridge, by design, displays the Gmail calendar on its screen display, to enable this, the user has to login to his Gmail account, and given the existing vulnerability in the smart fridge, the validation of the SSL certificate is failed and any hacker who manages to access the network on which the fridge is connected to, can potentially perform a Man in the Middle attack and steal the login credentials of Gmail account from the fridge calendar client. Therefore, any neighbor or person on the road outside, in theory, could steal the Gmail credentials of the user and hence compromise the privacy of the user [39].

However, this was not the first time when a smart fridge was compromised by hackers; between late 2013 and early 2014, the smart fridge was one among several smart gadgets involved in the kitchen and home media systems which were compromised by the hackers. It was estimated that more than 100,000 devices were compromised and were leveraged by the hackers to send spam mails as part of their junk mail campaigns, and the total number of messages sent crossed 750,000 [40].

Way back in 2000, Kaspersky Lab had foreseen this scenario and had warned that fridges and other appliances would be a potential target to malicious attackers and viruses could be designed in such a way that would enable the hacker to swing open a user's fridge at the middle of the night among other mischiefs.

With the increase in devices connected to the Internet, several vulnerabilities are introduced which become a threat to the security by creating avenues for all sorts of attacks. The poorly configured and appallingly protected IoT devices have raised concern among users who are transitioning to smart home appliances. Despite the early warnings, the persistence of flaws in smart devices only goes to show that security is still not a prime concern, and that enough testing is not being done before the smart devices are released. Many more of these attacks can be expected in the future if such a lax approach is continued toward security.

1.3.4.4 2017 Equifax Man in the Middle Attack

In March of 2017, one of the USA's major Credit Reporting Agency named Equifax was breached by an attack. Frequent attacks were done by the hackers. Approximately 145.5 million US consumers' personal data were stolen. More than 40 percent of the US population was affected. The compromised data mostly consisted of consumer names, addresses, Social Security Numbers (SSNs), License numbers, and Date of Birth. Information security experts were keeping an eye on dark websites after the attack to know if the data were leaked or not [41].

With the help of a consumer complaint web portal, the company was hacked. The attackers exploited the vulnerability which should have been patched by the organization. Vulnerability Apache Struts CVE-2017-5638, one of the popular frameworks used to create Java Web Applications was breached. This was a failure of Equifax's internal processes and a lack of judgment on their security analysts. Since the systems were not isolated from each other, attackers could easily move from a website to other servers. Due to this, they were successfully able to access further systems as they were able to find login credentials stored as plain text. Equifax had failed to renew an important certificate of encryption on one of their tools of internal security which was the reason behind the attackers pulling out data from the network for months (exactly 76 days) in encrypted form without being detected. The vulnerability was left unpatched from March 9, 2017, until July 29, 2017, when it was discovered by Equifax's Information security department. Equifax didn't disclose this breach for more than a month after discovering it had happened. The shareholders and top executives of the organization were accused of insider trading.

Similar attacks continued to impact Equifax affecting nearly 8,000 Canadian and 693,665 UK citizens. The web application was breached repeatedly, which tampered with the organization's image heavily. After thorough and extensive investigations, it was found that Chinese state-sponsored hackers were behind this with espionage being their intention rather than theft. This breach had cost Equifax more than $1.7 billion. Since this incident in 2017, Equifax ensured to have $125 million in cybersecurity insurance coverage [42].

The attackers were active within Equifax's networks for 76 days being unnoticed. If a user's data were compromised and leaked, then the maximum amount that could be asked for compensation was $125. Nearly $1.4 billion was spent immediately by Equifax for upgrading its security infrastructure. Before 2017, there hasn't been a single attack reported in Equifax.

The reason why the Chinese government sponsored this attack is that since 2015, similar kinds of attacks have taken place which attacked and got access to US citizens' data. This is a part of a mission to build a huge lake of data on millions of American citizens with the motive of using big data analytics to learn about the operatives of US government officials and intelligence. As per reports, nearly three attacks every minute takes place in the United States, which is a big concern as annually $3.8 million to $4 million is the average cost which the corporates have to bear. Also, various loopholes could be found which would be of great use to China. Retired or serving American officials or secret service agents who are facing monetary crunch could be bribed or blackmailed to become Chinese undercover spies. Recently, such events have taken place a lot which can't be a coincidence.

In February 2020, four members of the Chinese Armed Forces were charged in connection with the attack by the United States Department of Justice. This attack was taken seriously by the United States as they have rarely filed any criminal charges against officers of foreign intelligence.

Learnings to prevent such exploitations in the future are as follows [43]:

- Use some reputable sources such as MITRE to keep a track of Common Vulnerabilities and Exposures (CVE).
- A detailed list of all the libraries, dependencies of software, and key components (like Struts in Equifax Attack) which includes the last patch dates, versions, etc., must be kept.
- Spend nearly 20% of the budget to invest in security infrastructure which includes timely software patches and latest devices.
- Be sure to break down large programs into smaller ones so that you can manage these pieces independently and respond quickly to problems with the current version or CVEs, if any.

1.4 Limitations

There is a rush from industries to transition toward Industry 4.0, which would lead to higher productivity, quality, and flexibility, but the problem with this rush is that security is being undervalued and treated as a secondary requirement and the overall increase in cyber-based attacks on the industries is a clear indicator that there are certain vulnerabilities that need to be tackled and the severity of these vulnerabilities is under-appreciated in the industry.

The traditional architecture is being redesigned for the purpose of adapting to Industry 4.0, wherein a decentralized and autonomous model will be replacing the previously used master–slave architecture of the Computer-Integrated Manufacturing model (CIM).

Here, products and machines will be active participants in the IoT and regularly communicate with each other. Such an open environment is highly insecure and highly prone to both active as well as passive attacks and with Industry 4.0, the range of attacks is much higher whereas the possibility of detection is lower.

The biggest problem lies in the idea that cybersecurity is seen as a characteristic and not considered in the design phase; there is a misconception that cybersecurity can be purchased as a product or a service later on, and the existing threats are not well realized and often ignored as there are not too many reports from the industry, so there is a major misconception that hackers must not be advanced enough.

One key factor that separates, in particular, the smart manufacturing systems from the traditional IT systems is that the manufacturing systems are set up for the long haul and often have a lifespan of up to 20 years, most of the devices are legacy systems, and to keep these systems up to date, regular patching must be done, and herein lies a big problem.

For the systems to patch, the production must come to a halt, which leads to production downtime that incurs a loss, and demands additional costs; this is why the system owners are reluctant to patch frequently and it so happens, as with the Wannacry attack, that unpatched systems become vulnerable, and can be exploited [11,12].

Historically, we have seen that there are situations when systems are already exploited by the time a patch is released for an exposed vulnerability, and another issue is that expert skills are required to carry out the patching process.

The systems were designed for optimal performance but system security was not integrated with that optimization but rather was ignored and was planned to be incorporated as a service through vendors and thus a majority of the systems have only very little security like default passwords, no access control, and undocumented backdoors.

Potteiger et al. (2017) [44] have conducted tests on a platform that closely mimics the CPS and shown that the existing vulnerabilities in the railway transport network can cause a substantial amount of train delays whose effects propagate into the physical world and affect the various passengers involved. The researchers have performed a packet delay attack using a Man in the Middle attack and a combination of DoS with a Man in the Middle attack. For the first scenario where only the Man in the Middle attack was employed, the attackers were able to delay packets by 20 minutes causing delay to targeted trains, and on the other hand, when both Man in the Middle attacks as well as DoS attack was leveraged, the attackers were able to disrupt the entire functioning of the schedule of multiple trains (six in their example).

Humayed et al. (2017) [45] have identified the basic requirements of CPS security as data confidentiality, information integrity, availability, authentication, non-repudiation, and privacy protection. The researchers have outlined various threats to the network information layer, infrastructure security threats such as equipment damage or failure, and physical damage, and also laid out various information security threats spread across three aspects, namely physical layer threats, information layer threats, and application control layer threats.

Singh and Jain (2018) [46] have highlighted security and privacy as the concerning issues of CPS and also pointed out that managing big data produced by the interconnected devices in a secure manner is also a challenging task. Other challenges acknowledged by the authors were the security of devices and data transmission. The researchers have identified several vulnerabilities in CPS including vulnerabilities relating to network, platform, hardware, as well as software. Several technical as well as management vulnerabilities were also laid out. The several threats recognized by the authors were Malicious Software, Cryptographic Attacks, and Network Attacks, among others.

Nair et al. (2019) [47] have identified privacy, security, interoperability of systems, context awareness, and the design of the system as the challenges in Medical Cyber Physical Systems (MCPSs). The researchers have identified Data Breaches, Ransomware attacks, Social Engineering attacks, and insider attacks as the prominent threats in the field of MCPSs. They have also pointed out the difficulty in installing an IDS in MCPS actuator systems.

Systems are traditionally designed without incorporating security or by assuming that the system is isolated, hence safe. Secure software development process is not considered.

Poor security practices and policies are also a reason why vulnerabilities are introduced; the Industrial Control Systems make use of insecure communication protocols and they do not even have the capability to support authentication. The obvious and widely known good practices are not followed as one would normally expect.

There are generally two motives for attacks on these systems: exfiltration (i.e., breaching personal and/or sensitive data from compromised systems by sabotaging the normal operations of the systems) by collecting money and the other being political motive.

1.5 Security Measures

There are two types of security solutions existing at the moment, namely static and active defense.

To ensure static defense, proper and updated security regulations, standards, policies, and guidelines must be in place, and in order to ensure active defense, cryptographic techniques must be employed as devices in the CPSs are continually communicating with each other.

An Intrusion Detection and Prevention System must be in place which is able to effectively detect known attacks and at the same time possess the capability to perform intelligent analysis on suspicious traffic, and accurately detect malicious attacks and prevent them.

Eigner et al. (2016) [48] have devised an anomaly-based detection of attack approach which utilizes the normal system behavior in the ideal state as the training data and collates it with the ongoing system activity. This approach worked well in detecting Man in the Middle attacks as the system behavior was continually logged and when an attack took place, the operation was captured in the logging client, and the required features were extracted and analyzed. Finally, the unlabeled data were classified as an anomaly by detecting outliers. The model was able to classify the attack as an anomaly as there was extra transmission time incurred when the attacker intercepted the message and relayed the modified message, creating a pattern different from normal scenarios.

Gokarn et al. (2017) [49] have proposed an anomaly detection system to intensify the security of power grids by utilizing the Kalman filter. In their approach, they have used Power System Analysis Toolbox (PSAT) as the training data, which is fed to a SCADA parser and then given to the Kalman filter which produced estimated data which is then stored into a knowledge base, and in this way, a knowledge base is developed. For real-time scenarios, any incoming data is sent to the parser which passes it on to the Kalman filter, and the output of the filter is compared with the knowledge base to classify it as an anomaly, if an anomaly is found then an alarm is triggered. The real-time data also make their way into the knowledge base which leads to continual improvement of the classification model.

Potteiger et al. (2017) [44] have recommended the use of a platform which closely resembles the CPS and also incorporates a hardware in the loop so that there is a way to actuate and sense a physical system. With this testbed, the researchers were able to configure software with CPS and accurately analyze the network and software behavior with respect to the CPS. The proposed framework will be very useful as the design process of the CPS can be improved as per customized requirements of the CPS by analyzing the various threats with the help of the testbed with embedded hardware.

Wang et al. (2010) [50] have proposed a context-aware security framework for CPS using which several security objectives including encryption, authentication, key management, and so on can be incorporated with security-relevant context information; furthermore, context-incorporated communication protocols can be used to ensure increased safety.

1.6 Future Direction

1. **Improved Intrusion Detection System**: The presence of an IDS has been very useful in the CPS as it is able to detect known attacks with a high degree of accuracy courtesy of the predefined ruleset it follows; however, with the evolving nature of the cyber-attacks, there is now a need for an improved IDS, one which is able to intelligently detect anomalies with the help of Machine Learning and Deep Learning, which would ensure that any outliers from the normal operations are detected very early on, and a system to isolation should be available where the packets can be thoroughly analyzed in order to determine if it is a malicious packet. The ruleset should also be continually updated by leveraging the results obtained from training the models. An internal IDS is also needed in order to detect malicious activity going on inside the organization from any employee turned rogue.
2. **Learning from the IT Industry**: Although there are several underlying differences between the CPSs and the traditional IT Industry, it would serve CPS well if they were to ensure that security practices – which are followed by the IT industry like two-factor authentication, early detection and response models, cryptographic encryption of networks, segmentation of networks, etc., – are adopted in the CPS environment as well.
3. **Security Policy Creation and Enforcement**: Proper security policies should be in place at the CPS so that any possible mistakes are avoided as, historically, it has been found that there are huge losses accompanied by small mistakes from the employee's side; the organization must enforce these policies and ensure that proper compliance is there from everyone; this is very important as the human factor cannot be overlooked when thinking about a completely secure environment.
4. **Research**: The only way to stay on top of ever-evolving cyber threats is to always be up to date with the new cybersecurity practices and learn more about the upcoming threats, and quickly adopt the countermeasures against the emerging threats.
5. **Redesign the Design Phase**: It has been observed that security is not a primary concern when designing the CPSs, and the organizations instead plan to embed security at a later stage as a service, which historically has not served them well. If they could ensure security measures are well established right from the design phase, then lots of threats could be tackled without much hassle and eventual disastrous

consequences. For this to be successful, lots of testing need to take place by developing testbeds which have embedded hardware that would closely resemble the nature of CPS and conduct a thorough analysis of the system in order to find out the vulnerabilities that would arise, and take effective precautionary measures against them when designing the real CPS.

1.7 Conclusion

In this chapter, we have thoroughly examined the vulnerabilities that exist in the CPS and in the IT Industry in general, we have identified the drawbacks that exist in these systems and have considered several high-profile incidents in order to fully grasp the nature of the underlying threats of CPS. We have also identified the limitations in the CPS which industries must take care of and have provided various key security measures for that purpose, and, finally, we have suggested various ways using which the industries and future researchers could align their thinking so as to have a well-rounded secure environment.

References

1. R. Alguliyev, Y. Imamverdiyev, and L. Sukhostat, "Cyber-physical systems and their security issues," *Computers in Industry* vol. 100, pp. 212–223, Sep. 01, 2018. doi: 10.1016/j.compind.2018.04.017.
2. J. P. A. Yaacoub, O. Salman, H. N. Noura, N. Kaaniche, A. Chehab, and M. Malli, "Cyber-physical systems security: Limitations, issues and future trends," *Microprocessors and Microsystems*, vol. 77, Sep. 2020, doi: 10.1016/j.micpro.2020.103201.
3. A. Stefanov and C. C. Liu, "Cyber-physical system security and impact analysis," *IFAC Proceedings Volumes (IFAC-PapersOnline)*, vol. 19, pp. 11238–11243, 2014. doi: 10.3182/20140824-6-za-1003.00528.
4. L. J. Wells, J. A. Camelio, C. B. Williams, and J. White, "Cyber-physical security challenges in manufacturing systems," *Manufacturing Letters*, vol. 2, no. 2, pp. 74–77, Apr. 2014, doi: 10.1016/j.mfglet.2014.01.005.
5. R. Heartfield et al., "A taxonomy of cyber-physical threats and impact in the smart home," *Computers and Security*, vol. 78, pp. 398–428, Sep. 01, 2018. doi: 10.1016/j.cose.2018.07.011.
6. H. A. Kholidy, "Autonomous mitigation of cyber risks in the Cyber–Physical Systems," *Future Generation Computer Systems*, vol. 115, pp. 171–187, Feb. 2021, doi: 10.1016/j.future.2020.09.002.

7. J. Song and Y. Moon, "Security enhancement against insiders in cyber-manufacturing systems," vol. 48, pp. 864–872, 2020. doi: 10.1016/j.promfg.2020.05.124.
8. M. Z. Gunduz and R. Das, "Cyber-security on smart grid: Threats and potential solutions," *Computer Networks*, vol. 169, Mar. 2020, doi: 10.1016/j.comnet.2019.107094.
9. B. Mantha, B. García de Soto, and R. Karri, "Cyber security threat modeling in the AEC industry: An example for the commissioning of the built environment," *Sustainable Cities and Society*, vol. 66, Mar. 2021, doi: 10.1016/j.scs.2020.102682.
10. S. Phuyal, D. Bista, and R. Bista, "Challenges, opportunities and future directions of smart manufacturing: A state of art review," *Sustainable Futures*, vol. 2, Jan. 01, 2020. doi: 10.1016/j.sftr.2020.100023.
11. S.-C. Hsiao and D.-Y. Kao, "The static analysis of WannaCry ransomware," in *2018 20th International Conference on Advanced Communication Technology (ICACT)*, Feb. 2018, pp. 153–158. doi: 10.23919/ICACT.2018.8323680.
12. Frost Laurence and Tajitsu Naomi, "Renault-Nissan is resuming production after a global cyberattack caused stoppages at 5 plants," *Business Insider*, 15, May, 2017.
13. Pressley Alix, "Bouygues Group's construction subsidiary hit by Massive Ransomware Attack," *Intelligent CISO*, Feb. 19, 2020.
14. Brigs Bill, "Hackers hit Norsk Hydro with ransomware. the company responded with transparency," *Transform*, Dec. 16, 2019.
15. Hotter Andrea, "How the norsk hydro cyberattack unfolded," *AMM*, Aug. 22, 2019.
16. J. M. Hatfield, "Social engineering in cybersecurity: The evolution of a concept," *Computers and Security*, vol. 73, pp. 102–113, Mar. 2018, doi: 10.1016/j.cose.2017.10.008.
17. F. Salahdine and N. Kaabouch, "Social engineering attacks: A survey," *Future Internet*, vol. 11, no. 4. MDPI AG, 2019. doi: 10.3390/FI11040089.
18. F. Mouton, M. M. Malan, L. Leenen, and H. S. Venter, "Social engineering attack framework," Nov. 2014. doi: 10.1109/ISSA.2014.6950510.
19. Muncaster Phil, "Whaling spikes $2 billion in just two years," *Infosecurity Magazine*, Feb. 26, 2016.
20. Greenemeier Larry, "Massive insider breach at DuPont," Feb. 15, 2007.
21. Wilson Tim, "DuPont data thief sentenced to 18 months," *Dark Reading*, Oct. 07, 2007.
22. Lee Mathews, "Toyota parts supplier hit by $37 million email scam," *Forbes*, Oct. 01, 2019.
23. Korolov Maria, "Business email compromise attacks cost millions, losses doubling each year," *CSO*, Oct. 01, 2019.
24. Kottler Sam, "February 28th DDoS Incident Report," *The Github Blog*, Mar. 01, 2018.
25. Poremba Sue, "Types of DDoS attacks," *eSecurityPlanet*, May 04, 2017.
26. Porter Jon, "Amazon says it mitigated the largest DDoS attack ever recorded," *The Verge*, Jun. 18, 2020.
27. Ranger Steve, "GitHub hit with the largest DDoS attack ever seen," *ZDNet*, Mar. 01, 2018.
28. "Memcached DDoS Explained," *Akamai*, n.d.
29. Russell Jon, "The world's largest DDoS attack took GitHub offline for fewer than 10 minutes," *TechCrunch*, Mar. 02, 2018.

30. Swinhoe Dan, "Man-in-the-middle (MITM) attack definition and examples," *CSO*, Feb. 13, 2019.
31. Chivers Kyle, "What is a man-in-the-middle attack?," Mar. 26, 2020.
32. Vigliarolo Brandon, "Man-in-the-middle attacks: A cheat sheet," *Tech Republic*, Nov. 30, 2018.
33. "Cybersecurity History: The 1st Man-in-the-Middle Attack," *Havoc Shield*, Jul. 30, 2020.
34. "Man in the middle (MITM) attack," *Imperva*, Dec. 29, 2019.
35. Blackmore Nicole, "Fraudsters hacked emails to my solicitor and stole £340,000 from my property sale," *The Telegraph*, May 16, 2015.
36. "Fraudsters hacked emails to my solicitor and stole £340,000 from my property sale," *Cyber Security Review*, May 16, 2015.
37. Valia Hasti, "Protect your digital data from eavesdropping attacks," *Ilantus Technologies*, Nov. 22, 2019.
38. Leyden John, "Samsung smart fridge leaves Gmail logins open to attack," *The Register*, Aug. 24, 2015.
39. Venda Pedro, "Samsung smart fridge leaves Gmail logins open to attack," *Pen Test Partners*, Aug. 18, 2015.
40. "Fridge sends spam emails as attack hits Smart Gadgets," *BBC*, Jan. 17, 2014.
41. Fruhlinger Josh, "Equifax data breach FAQ: What happened, who was affected, what was the impact?," *CSO*, Feb. 12, 2020.
42. Lane Ben, "Equifax expects to pay out another $100 million for data breach," *HousingWire*, Feb. 14, 2020.
43. Price Ed, "The Equifax fallout: How organizations can prevent data breaches," *Devbridge*, n.d.
44. B. Potteiger, W. Emfinger, H. Neema, X. Koutosukos, C. Tang, and K. Stouffer, "Evaluating the Effects of Cyber-Attacks on Cyber Physical Systems using a Hardware-in-the-Loop Simulation Testbed," 2017.
45. A. Humayed, J. Lin, F. Li, and B. Luo, "Cyber-Physical Systems Security -- A Survey," Jan. 2017, Online.. Available: http://arxiv.org/abs/1701.04525
46. A. Singh and A. Jain, "Study of Cyber Attacks on Cyber-Physical System," *SSRN Electronic Journal*, May 2018, doi: 10.2139/ssrn.3170288.
47. M. M. Nair, A. K. Tyagi, and R. Goyal, "Medical Cyber Physical Systems and Its Issues," *Procedia Computer Science*, vol. 165, pp. 647–655, 2019. doi: 10.1016/j.procs.2020.01.059.
48. O. Eigner, P. Kreimel, and P. Tavolato, "Detection of man-in-the-middle attacks on industrial control networks," *Proceedings - 2016 International Conference on Software Security and Assurance, ICSSA 2016*, pp. 64–69, Feb. 2017. doi: 10.1109/ICSSA.2016.19.
49. V. Gokarn, V. Kulkarni, and P. Singh, "Enhancing cyber physical system security via anomaly detection using behaviour analysis," in *2017 International Conference on Wireless Communications, Signal Processing and Networking (WiSPNET)*, Mar. 2017, pp. 944–948. doi: 10.1109/WiSPNET.2017.8299901.
50. E. K. Wang, Y. Ye, X. Xu, S. M. Yiu, L. C. K. Hui, and K. P. Chow, "Security Issues and Challenges for Cyber Physical System," in *2010 IEEE/ACM Int'l Conference on Green Computing and Communications & Int'l Conference on Cyber, Physical and Social Computing*, Dec. 2010, pp. 733–738. doi: 10.1109/GreenCom-CPSCom.2010.36.

2

Applying Agile Practices in the Development of Industry 4.0 Applications

Carlos Ankora and D. Aju
Vellore Institute of Technology, Vellore, India

CONTENTS

DOI: 10.1201/9781003203087-2

2.1 Introduction

The technological revolution has already gone through three phases: the first, second, and third industrial revolutions. The fourth industrial revolution is still a growing trend within the technological environment. It is borne out of a convergence of information technology and operational technology, resulting in the conception of cyber-physical systems. The main elements driving Industry 4.0 are the connectivity among the systems to facilitate the exchange of data and the automation of the required processes leading to the optimisation and efficiency of the systems.

The technologies that facilitate Industry 4.0 include the Internet of Things (IoT), Big Data, Augmented Reality/Virtual Reality (AR/VR), Autonomous Robots, Cloud Computing, Cybersecurity and Simulation. Additional automation mainly characterises these technologies in the production processes and operations than the previous industrial revolutions. It connects the physical systems and digital technologies through cyber-physical systems. These systems can undergo configurations and alignment to fit the expected specifications and produce personalised decision-making results, indicating a transformation from mass production to customised production.

The chapter introduces the fourth industrial revolution and its technologies. Section 2.3 identifies challenges in the development of Industry 4.0 applications. Sections 2.4 and 2.5 focus on Agile Methodology and its application in Industry 4.0. Section 2.6 presents the Discussions, and Section 2.7 closes the chapter with the Conclusion.

2.2 Industry 4.0 Technology

The introduction of mechanisation, water and steam power engine in 1784 characterised the first industrial revolution. In 1870, the second industrial revolution followed, introducing electrical energy for mass production lines. Then followed the third industrial revolution in 1970 with digital, electronic and IT systems. The influx of digitisation and automation leads to the fourth industrial revolution, a convergence of information technology and operational technology that creates cyber-physical systems. The four industrial revolutions are represented in Figure 2.1.

The four main drivers of the fourth industrial revolution are the IoT, Industrial Internet of Things (IIoT), Cloud-based manufacturing and smart manufacturing. Industry 4.0 thrives upon nine pillars derived from these drivers shown in Figure 2.2. They include Big Data and Analytics,

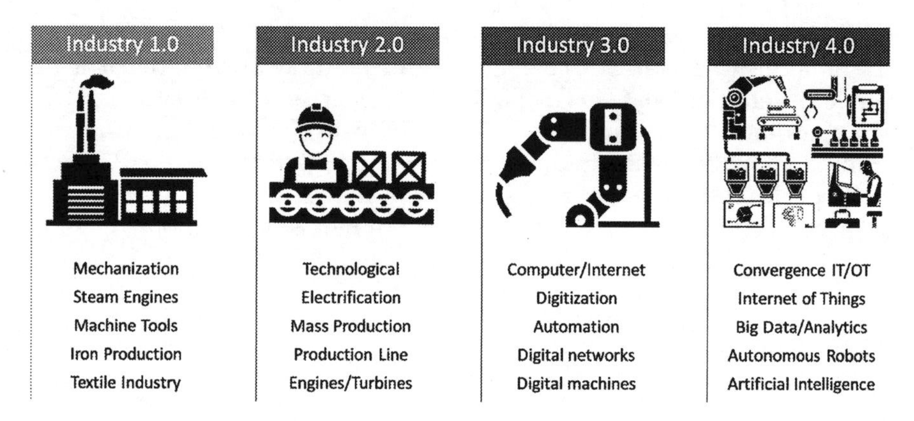

FIGURE 2.1
The first to fourth industrial revolutions.

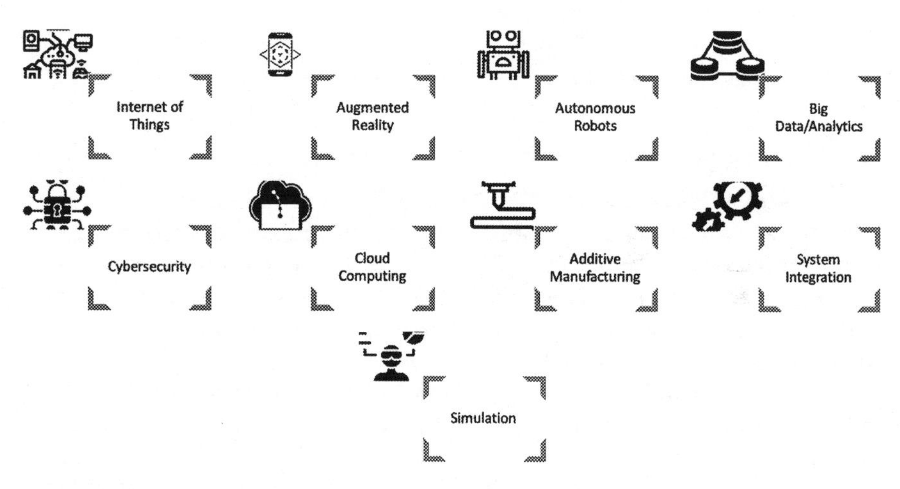

FIGURE 2.2
The nine technological pillars of Industry 4.0.

Autonomous Robots, Simulation, System Integration, the IoT, Cybersecurity, Cloud Computing, Additive Manufacturing and Augmented Reality [1].

Even though industry players and stakeholders have not fully exploited the technology, they have realised many benefits from its implementation. Automation of the processes in the development and manufacture of products improves productivity. The processes are also optimised, coupled with proper maintenance and evaluation measures resulting in higher business continuity. Monitoring takes place in real-time based on real-time data derived from the systems; this improves the supply chain and inventory processes. Industry 4.0 provides the opportunity to utilise organisational assets better, develop innovative capabilities and generate new revenue models. The consumers of the products are provided with personalisation and customisation preferences to improve the customer experience and have access to after-sales support and services.

The influx of e-commerce transactions has minimised the amount of interaction between customers and product vendors. Customers make online purchases all by themselves through an application interface, sometimes without any communication with the seller or manufacturer. Some customers these days prefer to incorporate customisation into their orders. This preference has made it necessary to consider the technology of smart industries and, by extension, smart services [2,3]. Automation of various tasks and processes improves the environment through the optimised use of machinery and resources and enhances safety in the operational environment. Overall, this improves business agility and provides better working conditions for the sustainability of business operations.

The pillars of Industry 4.0 creates the enablement for developers, users and customers to customise and utilise these technologies [4].

2.2.1 Internet of Things

The IoT sector has brought notable transformations in merging the physical and virtual worlds [4]. IoT enables connections among the various objects to transmit and exchange data. The network of components includes sensors, circuits, microcontrollers and other electronics embedded into home and office appliances, vehicles and other devices with the enablement of software and network connections. Many industries have implemented IoT, including healthcare, transportation, real estate, logistics and hospitality [5,6]. Industry 4.0, powered by the IoT and the Internet of services, with the connections of things, systems and users, has enabled the real-time transmission of data. The availability of necessary information helps in decision making and improves the product development process and the entire value chain [7].

The implementation of IoT comes with its challenges. One of the main issues is security threats and vulnerabilities, particularly privacy and denial of service. The involvement of various organisational units in IoT development and operations poses a challenge to the coordination and implementation of the right strategies leading to enhanced planning and efficiency [5,8,9]. Customers have embraced the new IoT technologies, which has led to changes in their needs and requirements. Several appliances and objects are getting digitised, and the demand is increasing. Manufacturing organisations should implement novel strategies and methods to develop IoT products and services to increase efficiency and productivity [6].

2.2.2 Augmented Reality/Virtual Reality

Augmented Reality and Virtual Reality blend the virtual and physical worlds, usually creating an immersive experience for users. During the development of Industry 4.0 applications, virtual objects could be shared among various interfaces, making the development process easier and more understandable [4]. In a mobile application utilising an Augmented Reality interface, the system communicates with the user via the device's camera's visualisation. The user also sends information back to the system, thus enabling a two-way interaction [10].

AR adds value to both the internal and external processes in organisational operations. AR promotes collaboration among the various units in the organisation's design process and process efficiency. AR also enhances customer and stakeholder experiences and the marketing activities to prospective customers [11–13]. The implementation of AR/VR in applications results in several benefits, including the reconfiguration of product designs to fit other objectives, visualisation of the configuration of products under development, enhancement of safety of production systems, better interaction between users and systems, providing remote support and conducting training within a safe environment to enhance productivity [14,15]. Augmented/Virtual Reality applications use several media types, text, audio, videos, animation and 2D and 3D models to provide rich content. They also provide users with an immersive experience creating a unique feeling of a surround virtual environment, enhancing the users' interaction and extending their engagement with the system [16,17]. Assistive systems implement AR/VR, providing users with self-support features and helping users with certain limitations perform tasks efficiently. One of the challenges with these systems is knowing the technology and how it operates. The inadequate understanding of the opportunities and constraints of these technologies and systems could limit the prospective users' and customers' appreciation and utilisation [18].

2.2.3 Big Data and Analytics

Big Data serves as an enormous collection of data gathered from various sources. It is mainly unstructured, comprising data from websites, blogs, chat forums, social media and application logs. Data users in organisations work with a wide variety of data. Database designers and data architects structure the data for storage; system developers use data to develop interfaces to accept, update and present data. End-users make data entries and run queries to generate reports. Customers prefer to have the correct detail of data about products to help make the purchasing decisions; managers prefer data in reports and visual presentations. Various techniques are employed to perform analysis on datasets. Data go through a mining process and is stored in a database using an SQL Query Engine. The users run ad hoc queries to generate reports and provide visualisations for presentation. Datasets have to be structured to serve the needs of all stakeholders, and that could be a challenge for Industry 4.0 [19–21].

Big Data has been relevant in the operations of various industries, including agriculture, education, sports, medicine and manufacturing [19]. Big Data gives data users a better and deeper perspective on the data. Data analytics help make decisions and improve the development and production process [4]. The right analytics could provide empirical domain knowledge about products and services and their production and development process leading to effective implementation of Industry 4.0 applications [22].

2.2.4 Autonomous Robots

Autonomous robots with the integration of the IoT are instrumental in developing self-directed and self-governing products through the production process in smart industries. They can develop options in situations that tend to lead to a deadlock. Within the manufacturing industry, autonomous robots oversee the operations of machines and detect faults and failures. They can perform diagnoses based on analytics from past data. In smart industries, reliance on robots (collaborative robots) is based on several issues, including the performance and safety of the robots, ease of use, flexibility and learning capability [4,23,24].

One of the challenges associated with autonomous robots is the management of the production workflow. Robots work with several movements of parts, accuracy and swiftness. Robotic applications should be continuously innovative, with new industrial processes and adaptation to novel conditions [23].

2.2.5 Cybersecurity

Security is of primary concern to organisations; they invest many resources in providing cybersecurity for their assets, including network communication systems, computer devices, software, data and other services. Attacks are launched on organisational systems to either destroy the system, deny access to computer systems and services, steal and give out information to unauthorised personnel or even sabotage the whole organisational infrastructure and devices [25,26].

One of the main threats to IoT systems is security. Some organisations find it challenging to hire experts who specialise in information technology and operational technology. However, it is necessary to have cross-functional teams that promote cross-functional learning [27].

Manufacturing systems need to be protected and secured from attacks. Addressing the cybersecurity issue in manufacturing applications requires the involvement of stakeholders [28].

Cybersecurity provides a prospect for customer value creation. Privacy concerns have become essential with Industry 4.0. Smart products are emerging, connecting devices, generating and storing loads of customer data; cybersecurity will concern customers and other stakeholders [29].

2.2.6 Cloud Computing

Cloud Computing provides the infrastructure for system users to have access to information and services without any geographical and time limitations; users can connect their devices irrespective of their operating platforms. Cloud-based manufacturing provides scalability, virtualisation, agility and networked manufacturing [30,31]. The IoT and Services generates data from the connectivity of sensors and other devices. The data need to be stored and analysed to make decisions and improve productivity. Cloud Computing integrates with IoT to provide business value to the manufacturing process. Analysed data from the systems and cloud services help reduce production costs and provide safety insights to support the production process and drive business and operational transformation [32].

With the enablement of cloud services, manufacturers can request requirements from customers and collect feedback about products and performance. Cloud Computing provides the platform for customers, suppliers and other stakeholders to collaborate and take decisions in product design, development, manufacturing, testing and all other activities involved in the lifecycle of products, including maintenance [30,33].

2.2.7 Additive Manufacturing

Additive Manufacturing, mainly associated with 3D printing, is the most disruptive technology in the Industry 4.0 transformation. In 3D printing, a 3D model is designed with computer-aided-design (CAD) software, and the model is printed from the 3D printer by adding layers of material to form a physical object. The materials include carbon fibres, thermoplastics, steel, biochemicals, metal and polycarbonate [34,35]. During the production of Additive Manufacturing, a study [36] proposes the characteristics of suitable products. The products should be customisable to a certain extent; they should be of low volume and products that can be optimised to give added functionality [36]. These characteristics make it easier to customise the 3D designs and models, and it is easier to make changes to the design as required. It is viable to produce the 3D physical objects in lesser batches. Several industries implement Additive Manufacturing, including food, fashion, electrical and electronics, biomedicine, pharmaceutical and casting [37].

A class of consumers who produce their goods developed; they are known as 'prosumers'. The customers have their requirements and preferences for products. These customers are involved in the production process and value chain from the onset. Customised products are thereby generated at a lesser time cost. It makes it easier for manufacturing companies to introduce new products because of the agility and flexibility in design [34,35,37].

2.2.8 Simulation

Manufacturing companies used to make productions and test the efficiency by experimenting with various methods until achieving a successful one. Manufacturers use virtualisation in the Industry 4.0 transformation to create model simulations and run tests on the simulations. Simulations help to improve product quality. Before a product is manufactured, the various components are simulated to evaluate their integration and how they interconnect. The simulations are tested to assess the efficiency and identify any possible risks before implementing the actual production [38,39].

Simulations are appropriate for conducting work training to ensure safety. It spans the production process, from designing models to testing and training. In situations where several scenarios are operated, simulation enables the different combinations to be experimented with to determine the most appropriate. Simulation combines with other Industry 4.0 technologies to increase efficiency. It helps to design connectivity among IoT devices; Additive Manufacturing creates 3D models as a form of simulation before the physical object is generated; Augmented Reality and Virtual Reality are forms of simulations of the natural environment [40,41].

The process of simulation could be involving and takes much time. Designing the models requires the input of experts and, more importantly, the users and stakeholders who know the domain. Users become more involved and possess a sense of ownership throughout the production process [39].

2.2.9 System Integration

System Integration, applied in Industry 4.0, is projected in horizontal, vertical and end-to-end. Horizontal integration could involve companies operating in different sectors, collaborating to achieve a common objective. Vertical integration operates within a company, creating integration among the various units in the company. End-to-end integration enables communication and interaction among virtual and physical objects as part of cyber-physical systems. These integrations facilitate product customisation and reduce operational costs by collaborating at the automation levels and integrating the entire value chain [38,42–45].

Among the levels of System Integration are coordination, cooperation and collaboration. These levels enable tasks to be conducted consistently and logically. The various components combine to attain a common objective of the system by cooperating and collaborating individually to form a connection. It creates the interconnection among the different entities of the system [42].

2.3 Challenges in Development

Customers of manufacturing products have been stunned by the level of digitisation embedded in the daily appliances used at homes, workplaces and other places. The demand for such products keeps increasing while customers continuously change their requirements. Production in Industry 4.0 is expected to be consistent, progressive, efficient and customer-centred. Innovative technological methodologies and practices are required to uncover new business opportunities [2,14,46].

During the development of systems, it is prudent to involve the stakeholders at the early stages of development. The stakeholders and consumers specify the requirements needed for the products. Sometimes, the requirements will be enormous and require prioritisation and scheduling because the development and production team cannot work on them within a limited timeframe. Industry 4.0 projects might be allocated limited resources, which should be managed well. The various stakeholders and partners would have to be involved in the development stages, requiring collaboration among them [47].

One major stakeholder during product development is the customer; they ought to be involved in the development and evaluation processes. Traditional development technologies have deficiencies in the integration of the customer, making them unable to have a thorough interaction and user experience of the new product features and functionality [48]. There is also the challenge of continuous systems integration and product delivery pipelines that are non-proprietary and enable the swift deployment of applications to IoT data sources.

One of the design principles proposed to help organisations develop approaches in implementing Industry 4.0 applications is modularity. When systems are developed in modules, it is efficient to maintain them when the requirements change. It is easier to plan for the unpredictable, especially in a dynamic environment where there are capabilities for swift changes. Modularity brings more flexibility and agility to the development environments [49,50]. Information transparency is also another one of the design principles. Industry 4.0, connecting technologies with devices and people, leads to enormous amounts of information required to take suitable vital decisions during the production and development process. Data includes the sensor and contextual data, virtual representations, simulation models, manufacturing and plant models and context-aware information. It is expected that the various stakeholders will make the necessary information accessible to promote transparency and enhance innovation and functionality [4,49,51].

2.4 Agile Methodology

The agile methodology is acknowledged as one of the development processes that emphasise user interactions to ensure maximum customer satisfaction. One of the practices that endear to agile development practitioners is the reliance on user feedback to improve both the development process and the product. Agile prioritises user interactions and customer collaborations over implementing tools and processes to develop software and products. Agile, through its practices, instils a particular mindset rather than following a set of methods. This mindset helps teams make better decisions to be more productive and deliver valuable products quickly [52,53]. Agile methodologies focus on responding to the continuously changing requirements and environment rather than being fixated on a laid-down plan.

Various system development teams implement agile practices. System development companies adopt agile methods to develop applications to be used by individuals and small and medium-sized organisations. Organisations that prefer bespoke systems consult system developers to provide solution needs. Teams heavily use agile methodologies as the stakeholders are involved in all stages of the development process and even form part of the development team. The focus has shifted from just delivering the final product to the customer to participating and getting innovative in the development cycle [4,54].

The standard agile practices applied during systems and product development include daily stand-ups, product backlog, sprint/iterations, sprint planning, sprint review, user stories, sprint retrospective, Kanban and pair programming. These practices allow cross-functional teams to self-organise and focus on delivering minimum viable products (MVPs) quickly. Changes and improvements to requirements are always welcomed, coupled with review sessions to improve the system and the development process. There are regular reflection sessions – daily and after each iteration when a value is delivered, to encourage transparency and team ownership.

2.4.1 User Stories

Scrum is one of the popular agile software development methodologies. Scrum practitioners employ several activities during system development; one of the most used practices is writing user stories. User stories are simple text used as part of the requirements elicitation process. The product owner usually writes the stories. The components include indicating the person (user) of the application, a functionality the user expects and why it is important to implement it in the application [55,56].

Developers collaborate in agile environment; all stakeholders are involved in the development process. Inadequate knowledge in the domain area of a system could affect the requirement elicitation process and result in deficiencies in the requirements specification. Developers and analysts conduct interviews, workshops, or questionnaires to identify the features and functionalities. They are then written on story cards in simple and atomic formats as part of a product backlog of features [57,58].

User stories are beneficial to agile teams as they promote collaboration and allow new and innovative ideas to be incorporated into systems. It helps to share information about the system's users, the tasks to be performed and the intended goal to be achieved [59,60].

2.4.2 Kanban

Kanban is a framework that originated from the manufacturing industry. It has become widely used in many industries, from factory floors to software development projects. Kanban is popular among agile development teams, but some non-agile teams also adopt Kanban to optimise workflow. Kanban comprises a three-step workflow – To Do, In Progress and Done. It utilises cards (or sticky notes) and a Kanban board; the backlog items are written on the cards and placed on boards. All team members can access the board and view other members' tasks. This structure helps limit work in progress and prevent overloading tasks. Kanban encourages transparency of work and enables easier collaboration among team members. Kanban helps teams identify bottlenecks in the workflow to improve the product development process [53,61]. Some of the foundational principles of Kanban include commencing with the current task and then continuously proceeding with new incremental tasks from the backlog. Kanban encourages transparency; the various roles, tasks and stages of development are visible to all, promoting collaboration and teamwork [52].

2.4.3 Sprint/Iteration

A sprint or iteration is a characteristic feature of Scrum. It denotes a designated period to work on a task, usually 2–4 weeks. Some teams vary the duration, preferring 1–3 weeks depending on the team's size, experience and scope of work. At the start of a sprint is a sprint planning session, where the team decides on the backlog tasks. At the end of the sprint, it is expected that the team will deliver an MVP, a potentially shippable product of value to the user. The team conducts a review at the end of a sprint to reflect on the development process and discuss improvements towards the next session. The team also provides input into the features for the product backlog. Sprint sessions have a fixed start and end date; the duration is maintained for all subsequent sprint sessions. Incomplete tasks during a sprint are added to the backlog and considered for the next sprint; the sprint duration is not extended to complete pending tasks [54,62,63].

2.4.4 Daily Stand-Up

Daily stand-up meetings characterise sprint or iteration sessions. The daily stand-up is a timeboxed meeting of about 15 minutes held daily among the agile team. Members conduct the meeting while standing to keep it brief and concise. Team members talk about what they have done since the previous meeting, open up on any hurdles or challenges they encountered in performing their tasks and inform other members on what they plan to work on during the day. There is much transparency involved; team members become aware of the level of progress and get the opportunity to make inputs to help with the team's progress. Teams use daily stand-ups to improve the developing products and the development process, ensuring checking up on other members' tasks, synchronising the roles and adapting to changes in requirements daily [52,54,63].

2.4.5 Sprint Review and Retrospective

A sprint review takes place at the end of a sprint session to reflect on the development process and discuss ways to improve in the next session. The agile team, including the stakeholders, developers, users and customers, is involved in the review process. The team provides input into the features for the product backlog. The last activity of the sprint is the retrospective. During the retrospective, the team discusses what went well and identifies what could be improved. They also discuss their commitments to improve the next sprint session. The Scrum master facilitates these sessions. The main focus is to inspect and adapt the product and the development process [54,63].

2.5 Agile Practices in Industry 4.0

The technologies of the nine pillars of Industry 4.0 endeavour to meet the continuously changing needs of the customer; by implementing agile development methodologies [4]. Cyber-physical systems are usually agile and necessitate evolving requirements for production to improve the systems continuously. Industry 4.0 incorporates several stakeholders to develop multi-dimensional solutions that satisfy their needs [64]. Production organisations usually focus on intermittent product releases and upgrades for customers. The situation has evolved with more iterative productions continuously incorporating changing customer needs [48]. Collaboration and communication are vital in meeting the customer's needs to improve the experience. The customer's involvement right from the inception of the development process through all the stages is crucial for the product's success. The customer can make valuable inputs and provide real-time feedback for integrating into the production [43]. Applying appropriate methods to elicit customer and user requirements is a significant aspect to consider to ensure an effective product development process [65].

During the development of Industry 4.0 applications, some steps have been identified to be followed [66]. The applications are expected to meet organisational goals and objectives. Requirements are elicited and gathered to start development; all relevant stakeholders should share a mutual interest and be clear on the requirements and objectives without any ambiguities. This clarity helps to be transparent and get all involved in the development process. It also helps to overcome any potential impediments right from the beginning, which is an excellent measure to deal with possible conflicts. There is a shared responsibility; however, the team members should understand their respective roles in getting tasks accomplished. Stakeholders ought to recognise the amount of effort, skill and experience they bring on board to achieve the ultimate goal of meeting the customer's needs.

Figure 2.3 represents an integration of the agile practices and the technological pillars of Industry 4.0. At the centre of the diagram is the

FIGURE 2.3
Applying agile practices on Industry 4.0 technologies.

customer who uses the applications developed from the technologies. The agile practices – user stories, Kanban, sprint, daily stand-ups and sprint review – help developers of the applications to involve and collaborate with the customer to create an enhanced customer experience. The following section presents the discussion on the diagram.

2.6 Discussion

In the current fourth industrial revolution, development and production are agile, mainly revolving around the customers and users to best meet their needs and continuously changing requirements. Manufacturers and producers of technological artefacts should invest in the digitisation and automation of processes, software, devices and training [4]. It is imperative to invest in the software and product development process to ensure prompt delivery of systems with value and improved functionality.

User stories are an excellent way to present requirements for stakeholders to understand better. In Augmented Reality/Virtual Reality systems, user stories will capture requirements and reflect on the storylines used in the application to create virtual objects and 3D models to augment the scene in the physical environment. In Autonomous Robots and IoT applications, user stories would highlight the main points of interactions that will take place in the cyber-physical systems to aid in the connectivity among devices. Big Data and Analytics aim to present data and reports so that the users can easily understand and help make the right decisions. Data visualisation helps tell a story about the data to be presented; the relevant story angles should be considered. User stories present the opportunity to structure data visualisations to tell stories in line with how the data users want it. During Simulations and Additive Manufacturing, user stories will describe how the physical object is supposed to look and how it will operate. The simulated designs and 3D models will then visualise the story before being produced as a physical object.

Software developers, product manufacturers and other industrial manufacturers use Kanban boards to visually monitor the workflow of tasks and activities. Kanban boards could be digitised or used physically. Tasks or story cards are placed on boards and segmented to indicate tasks yet to start, in-process and completed. They also help keep tasks within control while keeping limits for work in progress. For developing AR/VR applications, features are split into modules and tasks; Kanban boards are a good way of organising these tasks. The same applies to the workflow of autonomous robots and the IoT. When appropriately organised, it makes it easy to observe all the interconnections among the various objects and devices. Data analytics are performed to capture data across various organisational units. Kanban boards monitor the data representation among the various units visually. Simulated objects are split into several components and then integrated to evaluate the performance. Kanban could provide the visual workflow that organises each component into various developmental stages. Security is a never-ending operation. Each known cybersecurity threat and vulnerability could be placed on Kanban boards and observed by all team members to provide solutions. All the new threats are logged into the To-Do section and resolved threats are placed into the Done section.

The continuously evolving customer requirements are pushing manufacturers and developers to move away from providing one-time solutions to meet the needs of their clients, making the production and development process to be iterative. Agile methodologies like Scrum implement development in sprints. Each sprint is expected to deliver a viable product. When tasks are kept in a backlog, they are worked on during these sprints in order of priority. Data presentation and visualisation involve new aspects coming up each time. These new reports could be worked on during sprints. The same applies to the modules and scenes during Augmented/Virtual Reality application development; each module (or set of modules) could be worked on during each sprint, depending on the scope and amount of work involved. Additive Manufacturing involves creating 3D designs and printing them in layers. 3D printing could be done in various components completed in several iterations. Sprint sessions that involve teams from different functional units collaborate to work on features and products. The collaborative nature of sprints in a Scrum team would enhance the levels of System Integration.

During sprint sessions, the team conducts daily stand-up meetings to discuss members' challenges and consider tasks to be worked on during the day. Producing autonomous robots and cyber-physical systems requires technical expertise and collaboration among the team and other stakeholders. Daily stand-ups allow members to be transparent and develop a sense of ownership and trust in the project being undertaken. These practices help promote collaboration and teamwork, streamline the development process and the workflow, improve value delivery and encourage relentless improvement. Cybersecurity and Cloud Computing systems demand daily monitoring and operational activities. New threats and vulnerabilities emerge often, and issues of threats and compatibilities occur on cloud computing systems. Daily Stand-ups allow the team to reflect on previous challenges they faced and how to get them resolved as part of their daily tasks.

It is necessary to build cross-functional teams to collaborate during the development and production process. Review and Retrospective sessions are an effective way to reflect on what has been done and how to improve in future sessions. As requirements for Industry 4.0 applications continue changing, teams should continue to be iterative. Each sprint session should end in a review to consider improving subsequent sessions to create value for the customer.

2.7 Conclusion

The chapter primarily discussed the application of agile practices in the development of Industry 4.0 applications, spotlighting the benefits and challenges encountered during the development process. Industry 4.0 has presented customers the opportunity to demand personalised and customised products; this is relatively new to some manufacturers because of the continuous change in requirements. Technologies such as the IoT, Augmented/Virtual Reality, Artificial Intelligence, Autonomous Robots and Big Data and Analytics are key pillars in the digital revolution. Agile methodologies present practices that promote the involvement of customers and all relevant stakeholders in the software and product development process. Agile practitioners continue to promulgate the implementation of Scrum, user stories, Kanban, sprints, daily stand-ups and sprint reviews in system development. Industry 4.0 applications are customer-centric, aiming to meet the needs of customers and clients continuously; agile methodologies present the appropriate practices and approaches for developers and manufacturers to leverage on always to develop products that deliver value to meet the changing needs of the customer.

References

1. Vaidya S, Ambad P, Bhosle S. Industry 4.0 – A Glimpse. *Procedia Manuf* 2018;20:233–238. Available from: https://doi.org/10.1016/j.promfg.2018.02.034
2. Sharma A, Jain DK. Development of Industry 4.0. In: Nayyar A, Kumar A, editors. *A Roadmap to Industry 40: Smart Production, Sharp Business and Sustainable Development*. Cham: Springer International Publishing; 2020. pp. 23–38. Available from: https://doi.org/10.1007/978-3-030-14544-6_2
3. Stock T, Seliger G. Opportunities of sustainable manufacturing in industry 4.0. *Procedia CIRP* 2016;40:536–541. Available from: https://www.sciencedirect.com/science/article/pii/S221282711600144X
4. Kumar A, Nayyar A. si3-Industry: A Sustainable, Intelligent, Innovative, Internet-of-Things Industry. In: Nayyar A, Kumar A, editors. *A Roadmap to Industry 40: Smart Production, Sharp Business and Sustainable Development*. Cham: Springer International Publishing; 2020. p. 1–21. Available from: https://doi.org/10.1007/978-3-030-14544-6_1
5. Munirathinam S. *Industry 4.0: Industrial Internet of Things (IIOT)*. 1st ed. Vol. 117, Advances in Computers. Elsevier Inc.; 2020. p. 129–164. Available from: http://dx.doi.org/10.1016/bs.adcom.2019.10.010
6. Lampropoulos G, Siakas K, Anastasiadis T. Internet of things in the context of Industry 4.0: An overview. *Int J Entrep Knowl* 2019;7(1):4–19.

7. Koch V, Kuge S, Geissbauer R, Schrauf S. Industry 4.0—Opportunities and challenges of the industrial internet. *Fertility and Sterility* 2014. Available from: http://www.g20ys.org/upload/auto/03d10e20d681c1cfb40d3d9275704e2e9df16391.pdf
8. Küsters D, Praß N, Gloy Y-S. Textile learning factory 4.0—Preparing Germany's textile industry for the digital future. *Procedia Manuf* 2017;9:214–221. Available from: https://www.sciencedirect.com/science/article/pii/S2351978917301531
9. Bauer H, Baur C, Mohr D, Tschiesner A, Weskamp T, Alicke K, et al. Industry 4.0 after the initial hype: Where manufacturers are finding value and how they can best capture it. *McKinsey Digital* 2016. Available from: https://www.mckinsey.com/~/media/mckinsey/businessfunctions/mckinseydigital/ourinsights/gettingthemostoutofindustry40/mckinsey_industry_40_2016.ashx
10. Masood T, Egger J. Augmented reality in support of Industry 4.0—Implementation challenges and success factors. *Robot Comput Integr Manuf* 2019;58(February):181–195. Available from: https://doi.org/10.1016/j.rcim.2019.02.003
11. Mourtzis D, Zogopoulos V, Vlachou E. Augmented reality supported product design towards Industry 4.0: A teaching factory paradigm. *Procedia Manuf* 2018;23(2017):207–212. Available from: https://doi.org/10.1016/j.promfg.2018.04.018
12. Rauschnabel PA, Felix R, Hinsch C. Augmented reality marketing: How mobile AR-apps can improve brands through inspiration. *J Retail Consum Serv* 2019;49(March):43–53. Available from: https://doi.org/10.1016/j.jretconser.2019.03.004
13. van Lopik K, Sinclair M, Sharpe R, Conway P, West A. Developing augmented reality capabilities for industry 4.0 small enterprises: Lessons learnt from a content authoring case study. *Comput Ind* 2020;117:103208. Available from: https://doi.org/10.1016/j.compind.2020.103208
14. Damiani L, Demartini M, Guizzi G, Revetria R, Tonelli F. Augmented and virtual reality applications in industrial systems: A qualitative review towards the industry 4.0 era. *IFAC-PapersOnLine* 2018;51(11):624–630. Available from: https://doi.org/10.1016/j.ifacol.2018.08.388
15. Lavingia K, Tanwar S. Augmented Reality and Industry 4.0. In: Nayyar A, Kumar A, editors. *A Roadmap to Industry 40: Smart Production, Sharp Business and Sustainable Development*. Cham: Springer International Publishing; 2020. p. 143–155. Available from: https://doi.org/10.1007/978-3-030-14544-6_8
16. Aleksy M, Vartiainen E, Domova V, Naedele M. *Augmented reality for improved service delivery*. In: *2014 IEEE 28th International Conference on Advanced Information Networking and Applications*. 2014. p. 382–389.
17. Blanco-Novoa O, Fernandez-Carames TM, Fraga-Lamas P, Vilar-Montesinos MA. A practical evaluation of commercial industrial augmented reality systems in an Industry 4.0 shipyard. *IEEE Access* 2018;6:8201–8218.
18. Paelke V. *Augmented reality in the smart factory: Supporting workers in an industry 4.0. environment*. In: *19th IEEE International Conference Emergy Technology Fact Autom ETFA 2014*, 2014.
19. Sharma A, Pandey H. Big data and analytics in Industry 4.0. In: Nayyar A, Kumar A, editors. *A Roadmap to Industry 40: Smart Production, Sharp Business*

and Sustainable Development. Cham: Springer International Publishing; 2020. pp. 57–72. Available from: https://doi.org/10.1007/978-3-030-14544-6_4
20. Khan M, Wu X, Xu X, Dou W. Big data challenges and opportunities in the hype of Industry 4.0. *IEEE Int Conf Commun* 2017.
21. Santos MY, Oliveirae Sá J, Costa C, Galvão J, Andrade C, Martinho B, et al. A big data analytics architecture for industry 4.0. *Adv Intell Syst Comput* 2017;570:175–184.
22. Wang D. Building value in a world of technological change: Data analytics and Industry 4.0. *IEEE Eng Manag Rev* 2018;46(1):32–33.
23. Goel R, Gupta P. Robotics and Industry 4.0. In: Nayyar A, Kumar A, editors. *A Roadmap to Industry 40: Smart Production, Sharp Business and Sustainable Development*. Cham: Springer International Publishing; 2020. p. 157–169. Available from: https://doi.org/10.1007/978-3-030-14544-6_9
24. Indri M, Grau A, Ruderman M. Guest editorial special section on recent trends and developments in Industry 4.0 motivated robotic solutions. *IEEE Trans Ind Inf* 2018;14(4):1677–1680.
25. İlhan İ, Karaköse M. Requirement analysis for cybersecurity solutions in organisations. 2015.
26. Corallo A, Lazoi M, Lezzi M. Cybersecurity in the context of industry 4.0: A structured classification of critical assets and business impacts. *Comput Ind* 2020;114:103165. Available from: https://doi.org/10.1016/j.compind.2019.103165
27. Malatras A, Skouloudi C, Koukounas A. Industry 4.0 cybersecurity: Challenges & recommendations enisa lists high-level recommendations to different. 2019. Available from: https://www.enisa.europa.eu/publications/industry-4-0-cybersecurity-challenges-and-recommendations
28. Babiceanu RF, Seker R. Cybersecurity and Resilience Modelling for Software-Defined Networks-Based Manufacturing Applications. In: Borangiu T, Trentesaux D, Thomas A, Leitão P, Oliveira JB, editors. *Service Orientation in Holonic and Multi-Agent Manufacturing*. Cham: Springer International Publishing; 2017. p. 167–176.
29. Culot G, Fattori F, Podrecca M, Sartor M. Addressing Industry 4.0 cybersecurity challenges. *IEEE Eng Manag Rev* 2019;47(3):79–86.
30. Velásquez N, Estevez E, Pesado P. Cloud computing, big data and the Industry 4.0 reference architectures. *J Comput Sci Technol* 2018;18(3):e29.
31. Thames L, Schaefer D. Software-defined cloud manufacturing for Industry 4.0. *Procedia CIRP* 2016;52:12–17. Available from: http://dx.doi.org/10.1016/j.procir.2016.07.041
32. Georgakopoulos D, Jayaraman PP, Fazia M, Villari M, Ranjan R. Internet of things and edge cloud computing roadmap for manufacturing. *IEEE Cloud Comput* 2016;3(4):66–73.
33. Liu Y, Xu X. Industry 4.0 and cloud manufacturing: A comparative analysis. *J Manuf Sci Eng Trans ASME* 2017;139(3):1–8.
34. Baldassarre F, Ricciardi F. The additive manufacturing in the Industry 4.0 era: The case of an Italian FabLab. *J Emerg Trends Mark Manag* 2017;I(1):105–115. Available from: http://www.etimm.ase.ro/RePEc/aes/jetimm/2017/ETIMM_V01_2017_89.pdf

35. Haleem A, Javaid M. Additive manufacturing applications in Industry 4.0: A review. *J Ind Integr Manag* 2019;4(4):1930001.
36. Mellor S, Hao L, Zhang D. Additive manufacturing: A framework for implementation. *Int J Prod Econ* 2013;149:194–201.
37. Horst D, Duvoisin C, Vieira R. Additive manufacturing at Industry 4.0: A review. *Int J Eng Tech Res* 2018;8(8):3–8.
38. Alasdair Gilchrist. Introducing Industry 4.0 Industry. 2016; Available from: https://www.unido.org/fileadmin/user_media_upgrade/Resources/Publications/Unido_industry-4_NEW.pdf
39. Sturrock DT. *Traditional Simulation Applications in Industry 4.0*. 2019;39–54.
40. Gunal MM. *Simulation for the Better: The Future in Industry 4.0*. 2019;275–283.
41. Gunal MM. *Simulation and the Fourth Industrial Revolution*. 2019;1–17.
42. Sanchez M, Exposito E, Aguilar J. Industry 4.0: survey from a system integration perspective. *Int J Comput Integr Manuf* 2020;33(10–11):1017–1041. Available from: https://doi.org/10.1080/0951192X.2020.1775295
43. Pereira AC, Romero F. A review of the meanings and the implications of the Industry 4.0 concept. *Procedia Manuf* 2017;13:1206–1214. Available from: https://doi.org/10.1016/j.promfg.2017.09.032
44. Suri K, Cuccuru A, Cadavid J, Gerard S, Gaaloul W, Tata S. *Model-based development of modular complex systems for accomplishing system integration for Industry 4.0*. In: *Proceedings of the 5th International Conference on Model-Driven Engineering and Software Development*. Setubal, PRT: SCITEPRESS – Science and Technology Publications, Lda; 2017. p. 487–495. (MODELSWARD 2017). Available from: https://doi.org/10.5220/0006210504870495
45. Pisching MA, Pessoa MAO, Junqueira F, dos Santos Filho DJ, Miyagi PE. An architecture based on RAMI 4.0 to discover equipment to process operations required by products. *Comput Ind Eng* 2018;125:574–591. Available from: https://www.sciencedirect.com/science/article/pii/S0360835217306034
46. Lee J, Kao HA, Yang S. Service innovation and smart analytics for Industry 4.0 and big data environment. *Procedia CIRP* 2014;16:3–8. Available from: http://dx.doi.org/10.1016/j.procir.2014.02.001
47. Sarvari PA, Ustundag A, Cevikcan E, Kaya I, Cebi S. *Technology Roadmap for Industry 4.0*. Cham: Springer, 2018;95–103.
48. Nunes ML, Pereira AC, Alves AC. Smart products development approaches for Industry 4.0. *Procedia Manuf* 2017;13:1215–1222. Available from: https://doi.org/10.1016/j.promfg.2017.09.035
49. Hermann M, Pentek T, Otto B. *Design principles for Industrie 4.0 scenarios*. In: *2016 49th Hawaii International Conference on System Sciences (HICSS)*. 2016. p. 3928–3937.
50. Gattullo M, Scurati GW, Fiorentino M, Uva AE, Ferrise F, Bordegoni M. Towards augmented reality manuals for industry 4.0: A methodology. *Robot Comput Integr Manuf* 2019;56(October):276–286. Available from: https://doi.org/10.1016/j.rcim.2018.10.001
51. Helmold M. Industry 4.0 and Artificial Intelligence (AI) in PM. 2019. p. 161–163.
52. Stellman A, Greene J. Learning Agile – Understanding Scrum, XP, Lean and Kanban. *J Chem Inf Model* 2015; 53:1689–1699.

53. Al-Zewairi M, Biltawi M, Etaiwi W, Shaout A. Agile software development methodologies: Survey of surveys. *J Comput Commun* 2017;5(5):74–97.
54. Sommerville I. *Software Engineering*. 10th ed. Pearson; 2015.
55. Lucassen G, Dalpiaz F, van der Werf JMEM, Brinkkemper S. Improving agile requirements: The quality user story framework and tool. *Requir Eng* 2016;21(3):383–403.
56. Lucassen G, Dalpiaz F, van der Werf JMEM, Brinkkemper S. *The use and effectiveness of user stories in practice*. In: *International Working Conference on Requirements Engineering: Foundation for Software Quality*, 2016;205–222.
57. Bolloju N, Gupta A, Alter S, Gupta S, Jain S. *Improving scrum user stories and product backlog using work system snapshots*. In: *AMCIS 2017 – America's Conference on Information Systems: A Tradition of Innovation*. 2017. pp. 1–10.
58. Inayat I, Moraes L, Daneva M, Salim SS. *A reflection on agile requirements engineering: Solutions brought and challenges posed*. In: *ACM International Conference Proceeding Series*. 2015.
59. Quesenbery W, Brooks K. *Storytelling for User Experience: Crafting Stories for Better Design*. Rosenfeld Media. 2010.
60. Forbrig P. *Use cases, User stories and BizDevOps*. In: *Jt Proc REFSQ-2018 Work Dr Symp Live Stud Track, Poster Track co-located with 23rd Int Conf Requir Eng Found Softw Qual (REFSQ 2018)*, Utrecht, Netherlands. 2018; 2075.
61. Poppendieck M, Poppendieck T. Lean software development: An agile toolkit. *Computer* 2003;36.
62. Srivastava A, Bhardwaj S, Saraswat S. *SCRUM model for agile methodology*. In: *Proceeding – IEEE Int Conf Comput Commun Autom ICCCA 2017*, January 2017; 864–869.
63. Rubin KS. *Essential Scrum: A Practical Guide to the Most Popular Agile Process*. 1st ed. Addison-Wesley Professional; 2012.
64. Lu Y. Industry 4.0: A survey on technologies, applications and open research issues. *J Ind Inf Integr* 2017;6:1–10. Available from: http://dx.doi.org/10.1016/j.jii.2017.04.005
65. Engelbrektsson P, Söderman M. The use and perception of methods and product representations in product development: A survey of Swedish industry. *J Eng Des* 2004;15(2):141–154. Available from: https://doi.org/10.1080/09544820310001641245
66. Kumar A, Gupta D. Challenges within the Industry 4.0 Setup. In: Nayyar A, Kumar A, editors. *A Roadmap to Industry 40: Smart Production, Sharp Business and Sustainable Development*. Cham: Springer International Publishing; 2020. p. 187–205. Available from: https://doi.org/10.1007/978-3-030-14544-6_11

3

A Review of Security Vulnerabilities in Industry 4.0 Application and the Possible Solutions Using Blockchain

Mangayarkarasi Ramaiah, Vasavi Chithanuru and Adla Padma
Vellore Institute of Technology, Vellore, India

Vinayakumar Ravi
Prince Mohammad Bin Fahd University, Khobar, Saudi Arabia

CONTENTS

DOI: 10.1201/9781003203087-3

3.1 Introduction

The term "Industry 1.0" was coined in the 18th century by exploiting the steam power and automation of production systems. Industry 2.0 was figured out in the 19th century, benefited through the invention of electricity and assembly line production, and its process greatly improvises the automation compared to its earlier version. Industry 3.0 version introduces semi-automation in the manufacturing sectors. Robots were used to ease the human burden on many tasks. Industry 4.0 is meant for manufacturing sectors that aspire to remove the hurdles faced in collaborating and sharing data and information (Erboz, 2017). Industry 4.0 enhances the process and modalities to be followed in data exchange while implementing automation in the manufacturing industry. One of the crucial parts of Industry 4.0 is guaranteeing cost-effectiveness in the assembling areas. Industry 4.0 comprises the following sectors: Industrial Internet of Things (IIoT), IoT, Smart factories, AI and Cloud Computing, and Big Data. The entire device components meant for the automation are connected and interact with humans. Such an interconnected setup is termed Cyber-Physical System (CPS). The cyber component is associated with the computer-based algorithm to control as well as monitor the process. Table 3.1 displays the evolutionary version of the Industrial Revolution (IR) from 1.0 to 4.0 along with its drawbacks and the same has been presented graphically in Figure 3.1.

As the entire assembling measure is on mechanization guaranteeing, security is a significant viewpoint for businesses. The core design aspects of Industry 4.0 are solving compatibility issues, decentralization, virtualization, and transparency. While implementing the core design aspects, a lack of knowledge about modalities in setting up the environment creates most of the security vulnerabilities (Anisetti et al., 2020). As indicated by

TABLE 3.1

Evolutionary Versions of the Industrial Revolution

Industrial Revolution	Cryptonym	Established	Demerits
Industry 1.0	Power Generation	1784	Causes environmental pollution
Industry 2.0	Industrialization	1870	Framing the assembly lines for electrification
Industry 3.0	Electronic Automation	1969	Human intervention is essential
Industry 4.0	Smart Automation	2011	More challenging technology is required to protect from cyber threats

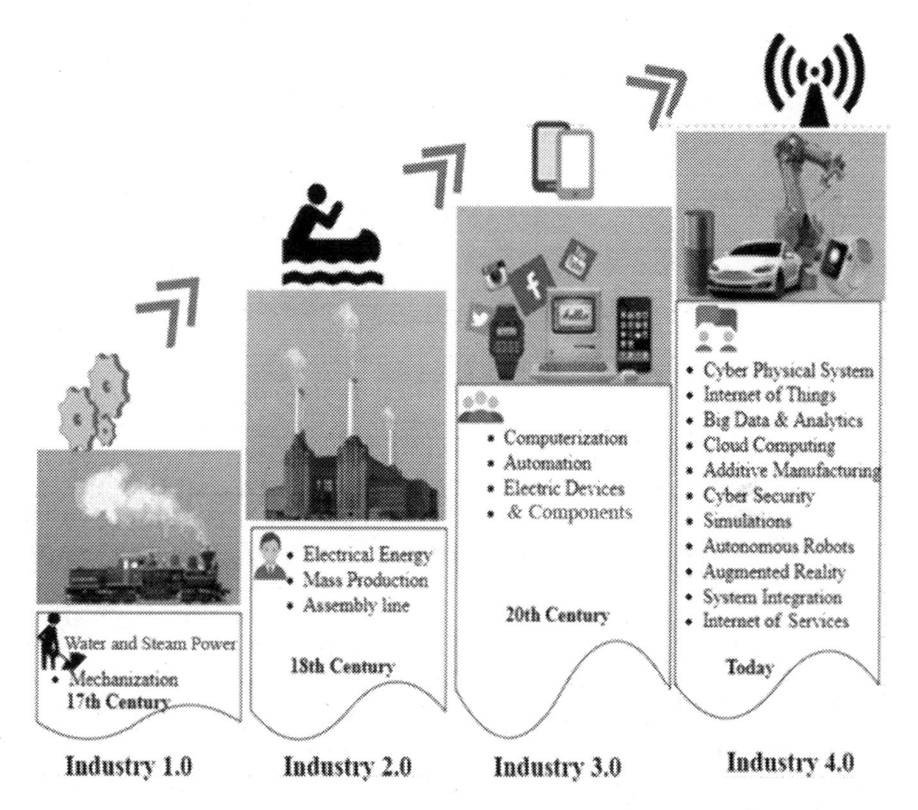

FIGURE 3.1
Industrial Revolutions 1.0 to 4.0.

IBM X-Force Threat Intelligence Index, 2018, 33% of the relative multitude of attacks rose against the manufacturing sectors. In a smart manufacturing environment, the furthermost regular attacks are Denial of Service (DoS) Attacks, Malware threats, Man in the Middle (MITM) Attacks, and Side-Channel Attacks.

Nowadays, the primary challenge for Industry 4.0 is ensuring security between its incompatible key components. Naturally, interconnected devices are facing difficulties in ensuring privacy and trust. Ensuring privacy prevents unauthorized access to confidential information. To include privacy, cryptographic algorithms ECDSA (Elliptic Curve Digital Signature Algorithm), MPC (Multi-Party Computation), and lightweight cryptographic algorithms are used. The other security aspect is where the organization finds it difficult to prevent vulnerabilities in the action that happened by compromising integrity. Hashing functions are facilitated with the system well in terms of the trust aspect. Apart from these, to establish trust between the people, we use different types of Models. The possibilities of data breaches are crippling the product deployment process before they gained revenues and adversely affecting the industrial economy (Chiu et al., 2017). In the context of IR 4.0, any piece of information is valuable and is subject to being hacked as well. Industry giants Centrify and Palo Alto Networks, implement a zero-trust policy in contradiction to privileged access methods. Conventional privileged access methods are not sufficient for the current trend, where, advanced persistent threats are surging. The privileged access method not only needs to cover databases and interconnected devices but also needs to be expanded to Big Data, cloud environments, and automated applications. The main idea of Zero trust privilege is permitting the user with the least privileges instead of ignoring them. The user request can be verified upon the requesting access, the context of the request, and the potential risk of the computing environment. Through least privilege access, the possible attacks on the industry operation can be minimal to a certain extent, improve audit and reduce the complexity and costs for the enterprise. Consequently, ensuring security aspects for IR 4.0 is vital to improvise the operation in various perceptions. Thus, the proposed chapter attempts to summarize the potential vulnerabilities due to compromising the various security measures and possible solutions to mitigate the threats using Blockchain Technology (BCT). Section 3.2 converses the several key components of Industry 4.0. Section 3.3 summarizes the inherent challenges from each key component wherever possible. In addition, the possible security challenges anticipated in every component are also explored. Section 3.4 discusses the various state-of-the-art industry applications intended to improvising the quality of human life ahead of smart city migration. Section 3.5 discusses how BCT has been used to increase the security measures of various applications in Industry 4.0. Section 3.6 concludes the chapter.

3.2 Components of Industry 4.0

Industry 4.0 applications are built with robust trending technologies to face challenging security attacks as well as recognize legitimate entries. This section discusses the main components of Industry 4.0 with the most probable events of security breaches. Figure 3.2 shows the various components of Industry 4.0, which are as follows: CPSs, Machine to Machine (M2M), Internet of Things (IoT), Big Data and Analytics (BDA), Cloud Computing (CC), Edge Computing (EC), Cyber Security (CS), Internet of Services (IoS), System Integration, Autonomous Robots, Simulations, Augmented Reality (AR), and Additive Manufacturing.

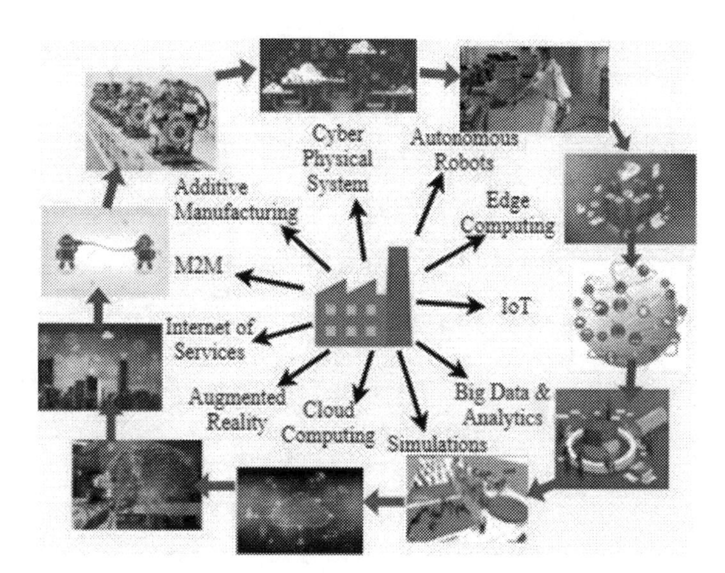

FIGURE 3.2
Components of Industry 4.0.

3.2.1 CPS and IoT

Industry 4.0 offers the merits of futuristic technologies and the manufacturing industry. Oztemel and Gursev (2020) say that it is the integration of computing and physical components. The development of a CPS is categorized into three phases.

- CPS incorporates identification advances like RFID labels that permit extraordinary recognizable proof. Capacity and examination are ought to be given as a focal help.
- For specific sensors and actuators, the CPS is equipped with fewer functionalities.
- CPS is designed with many sensors and actuators to become compatible with the network.

The IoT includes data in silos and information between millions of devices focussed on the protocols for managing the devices in terms of their configuration setup, platforms, etc. Tay et al. (2018) describe Industry 4.0 as the new period which embraces the advances like the IoT and the manufacturing industry. Georgios et al. (2019) disclose that by executing and adjusting to Industry 4.0 and IoT advances, extraordinary degrees of financial development and usefulness proficiency can be accomplished by enterprises and many as follows:

- Improvement of general application, administration, and framework accessibility and viability.
- Reduced time-to-market due to increased efficiency and shorter lead times.
- The assistance of the variation to individualized client necessities and market requests.
- Improvement of observing and controlling enterprise's cycles and resources.
- Reorganization and making digital production.
- Ability to examine a large amount of data within a short span of time.

3.2.2 Big Data and Analytics

Big Data offers a massive amount of storage for all industry applications. Big Data analytics comprises the merit mechanisms to handle huge amounts of data and Artificial learning techniques, thereby promoting insightful data analysis. BDA is designed to offer authentic data on a potential asset previously utilized in the creation of a similar descent, which might stay reprocessed. AI systems are the most promising platform to explore and investigate a huge amount of data for precise forecasting. Erboz (2017) describes that BDA is able to store and break down all the business information. Its benefit for Environmental Management is to bunch, break down, and give significant information and data on existing properties, properties utilized, machine usefulness, energy effectiveness, waste generation, waste utilization, and contamination levels transmitted.

3.2.3 Cloud Computing

Cloud Computing is a framework where it's anything but a gigantic measure of capacity for the client. Cloud Computing innovation is to be approved inside the manufacturing and individual administration exercises (Erboz, 2017). The exploration of an innovative solution is always a continuous process, in any context, and, therefore, data storage is designed to be in a distributed manner, allowing the creation of structures to be more data-driven. Cloud Computing is an excellent platform for Big Data preferred by numerous organizations during their information framework fabrication (Bajic et al., 2020). The objective of Cloud Computing is an accentuation on BDA that should utilize hardware equipment to fabricate figuring bunches and scale out the registering limit concerning web slithering and ordering framework jobs. Because of the gigantic volume of a dataset, looking for an ideal arrangement with adaptation to non-critical failure computational limit is a significant factor for executing Cloud Computing for BDA. Cloud Computing offers a computing service where data shared and processed through scalable resources are available on the Internet. To circulate their work, computer assets are put in numerous areas where these software parts have arrived behind schedule in a bunch. This strategy is utilized for making an examination that runs all the more quickly and is fit for playing out the tedious and power-consuming information collection.

3.2.4 Edge Computing

Edge Computing is a circulated organizing design in moving the computational force of the cloud closer to information age sources to diminish idleness and cut information-moving costs (Oztemel & Gursev, 2020). Modern edge computing is the portrayal of consolidating edge processing innovations such as in manufacturing and production environments (Sittón-Candanedo et al., 2019). Edge computing ensures the successful implementation of Industry 4.0 by leveraging distributed networks. A few of its merits are summarized as follows.

- The unworthy data streamed from various sources are filtered to save time and storage capacity.
- The aspect of scalability is greatly enhanced with decentralized storage and processing.
- The hubs of Edge models furnish every hub of the organization with isolation and security.
- Low latency is achieved by greatly reducing the data transformation.

3.2.5 Internet of Services

The Web of Services plays a significant role in the automotive industry. It goes about as "service vendors" to offer types of assistance through the web as indicated by the kinds of digitalization administrations. These administrations are accessible and on request around plans of action, accomplices, and any arrangement for administrations. In Industry 4.0, in the term "Internet of Things", "Things" represent the nuts and bolts meant for the process – the highlight of automation implemented through various services. Industry 4.0 consists of both hardware and software. Many organizations release their software updates online. Therefore, through the Internet of Services, the software of any component can be updated. Some of them generate revenue for the software update as well. Such kinds of services greatly support the Industry applications.

3.2.6 Autonomous Robot, Simulation, and Augmented Reality

Autonomous robots are designed with futuristic technologies capable of doing a task on their own without relying on human intervention. Robots find their applications in various sectors of automation Industries to facilitate the tasks involved in supply chain management, smart factories, and mobile robots, which are as follows. Robotics deployment ensures overhead cost reduction, minimal human error, and hence obtain complacency in production. Recently, mobile robots have been playing a role in alleviating the hindrances involved in the logistics of manufacturing, such as shifting materials by navigating to different areas. Hence, mobile robots reap merits such as flexibility, scalability, and safety. Tay et al. (2018) believe that the inherent adaptability of Autonomous Robots makes them suitable to assist in many fields. Later on, robots are designed to interact with their peers to perform group activities, including humans. Simulation is a significant apparatus and is additionally perhaps the most mainstream part of Industry 4.0 and reformed in the operational cycles through increased reality. Simulation models are the best tool to measure the efficacy of the project intended for robots and Augmented Reality. The virtual arrangement of various components doesn't necessitate cost; it simply reduces the complexities anticipated in a real environment. Hence, administrators can further develop the machine settings in an essentially replicated situation prior to executing them in the real context. Thus applying simulation modeling for Industry 4.0 enhances productivity in product delivery, supply chain management, and manufacturing sectors. AR considers the objects from the real world and the virtual world. In trend, industries are looking forward to the possibility of an innovative solution to augment their production. AR offers many solutions to the industries through the way information can be accessed or analyzed especially in smart factories. Carvalho et al. (2020) present that while creating smart manufacturing functionalities, Augmented Reality is one of the technological advancements intended for Industry 4.0. Though a couple of years prior, Augmented Reality was simply seen as an extravagant toy, now it has arrived at the development level utilized in a production environment, where AR supports executing the design plan virtually before investing the money for the necessary sources. During the simulation, they can see the possible error-prone zone, so before inserting the component in real-time, bugs can be fixed.

Hence, AR helps the manufacturing sectors in speeding up the operation with minimal error and thereby facilitating economic growth.

3.2.7 Additive Manufacturing and Machine to Machine

Industry 4.0 stresses the importance of the integration of smart technologies and the production system. According to Oztemel and Gursev (2020), the additive manufacturing (AM) technique includes a variety of composite geometries and designs from three-dimensional (3D) model information. Industry 4.0 is invigorating the tradition of cutting-edge data advances and keen creation frameworks. The process comprises printing progressive physical layers that are enclosed on top of one another. Industry 4.0 execution relies upon added substance-producing abilities. The AM contains changed methods, materials, and equipment, has been established all over the long term, and can modify gathering and coordination actions. The AM has been generally functional in a variety of schemes with biomechanics, prototyping, and development. This innovation utilizes various materials for various purposes, for which viewpoints like limit, strength, consistency, limits, and contact with other potential sorts of materials should be noticed. M2M is direct communication between tools using any channel, wired or remote. M2M communication can include current instrumentation, allowing a sensor or meter to take the material where it archives to application programming that can apply it. In Industry 4.0, organizations utilize the M2M innovation to build up remote correspondence between data focuses and machines. Making correspondence advances, link or remote, simple to execute, and as modest as conceivable has opened up the developments for a simpler life. As to Industry 4.0, M2M is additionally viewed as a fundamental part.

3.3 Inherent Challenges for Industry 4.0 Components

In Section 3.2, the numerous important mechanisms of Industry 4.0 were discussed which portrays the ideas of various authors. Uninterrupted data exchange through the secured communication channel is mandatory to accomplish the task without compromising the security aspects. The Industry 4.0 environment is more prone to be hacked because of the nature of the components involved in it; the main intention of including this section is to familiarize the readers with the most possible security challenges (mentioned in Figure 3.3) that may be anticipated from Industry 4.0 components.

3.3.1 Cyber-Physical System, IIoT, and Cyber Security

A CPS comprises many computer systems where the candidate process is controlled and monitored through intelligent steps. And, it has become one of the essential construction components for the trending Industries. Various benchmarking procedures are presented in various venues to enhance their efficacy as well as to enrich security measures.

To communicate heterogeneous components and adaptive support for components, there is a great requirement for a flexible interface for

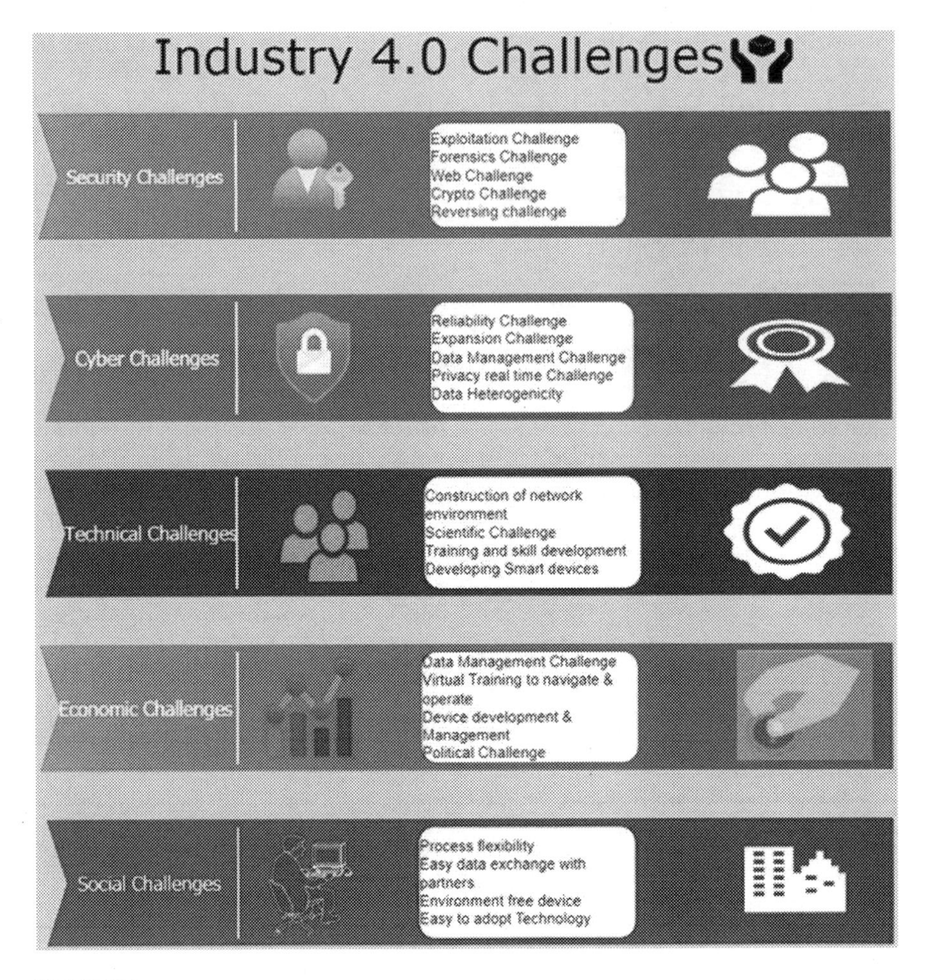

FIGURE 3.3
Industry 4.0 challenges.

communication. Extracting the common features of components from the CPS file is challenging because each file contains methods, languages, and models (Zhou et al., 2015), the design and development of the CPS tool support synthesis, simulation, improvement, sculpture, and simulation. Bajic et al. (2020) claimed that a huge investment in terms of cost is required to develop a CPS technology and also train the employees due to a lack of knowledge in setting up the various components. Zhang et al. (2019) summarize the possible integration challenges that may be expected from CPS. Keerthi et al. (2017) demonstrate the security challenges while ensuring the security aspects such as Trustworthiness, Exactness, Efficiency, Robustness, Versatility, and Security along with the possible solutions. Security breaches may be higher for interconnected devices such as CPSs. The following are the most anticipated type of security breaches for the CPS: Sybil attacks, Eavesdropping, Packet Spoofing, and Feedback Integrity attacks. The security breaches of the IoT are Data transit attacks, Jamming DoS attacks, Routing & DoS attacks, and Unfairness attacks.

The components, the IoT and IIoT are very essential nowadays to facilitate many transactions between the industries (Serror et al., 2020). During the communication between the interconnected devices in Industry 4.0, there are a lot of security breaches that may be anticipated in the IoT and IIoT. Some of these may be physical breaches, impersonation breaches, MITM, routing breaches, malicious code injection vulnerabilities like DDoS, crypto-jacking breaches, ransomware, data leakage, DoS, passive and active attacks, and inside and outside attacks.

- The physical breach is one of the categories where the problems are faced due to hardware components in IoT. That is, the device is going to be accessed physically and replaces or damages the components to access the sensitive information like passwords of users or by accessing the device.
- Impersonation breaches can be succeeded with a poor authentication mechanism.
- MITM is one of the breaches that focuses on the information shared between any two legitimate parties.
- While sending the message from one partner to another, the attacker impends the communication network by broadcasting false routing information or by disturbing the flow of communication in the Routing breaches. Examples of this type are Sinkhole and selective forward.
- Distributed DoS is one of the problems in Industry 4.0 because large IoT-based devices are used. There is a chance of attacking a server by congesting with requests.
- In a crypto-jacking breach, an unauthorized person uses computational properties to mine a cryptocurrency.
- Ransomware demands money from the user by blocking content visibility.
- Leakage of data is one of the security breaches for compromising personal information.

DoS attacks show a key part in IoT devices. These attacks faced problems by overloading the machines with requests.

The security breaches of the IIoT are brute force attacks, SQL injection, mobile devices attacks, MITM attacks, web application attacks, replay

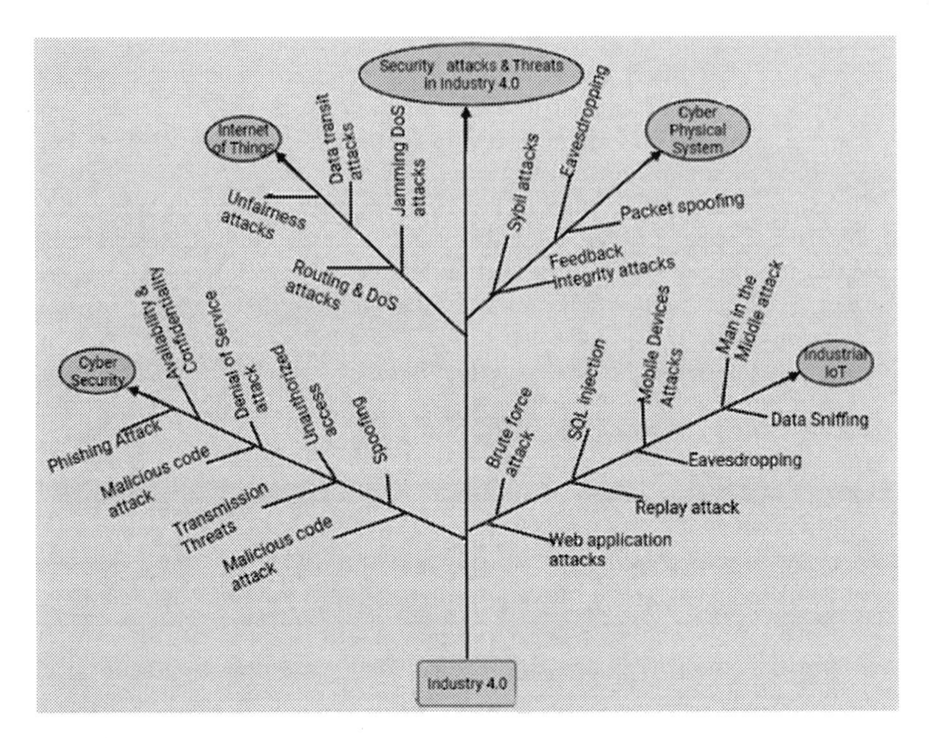

FIGURE 3.4
Most possible security breaches for the components of Industry 4.0.

attacks, and data sniffing. Security breaches of Cyber Security are Phishing attacks, malicious code attacks, DoS attacks, transmission threats, unauthorized access, and spoofing. The most anticipated security breaches for the components in Industry 4.0 are pictured in Figure 3.4.

3.3.2 Cyber Security

Cyber security deals with the security aspects of devices connected in networks of networks. According to Kumar et al. (2020), the likelihood of Phishing Attacks is more while surfing using the Internet. Phishing attacks are designed to steal user's sensitive information. Such attacks are triggered when the user enters user credential information or downloads some pop-ups. Phishers utilize the gap in compromised attacks, which consolidates social designing, focusing on both the absence of specific dynamic safety efforts by frameworks and the absence of data or watchfulness of clients. This includes attacks such as zero-day malware, connect control, channel avoidance, jumbling brand logos, site falsification, secretive divert, and break of IIoT frameworks and, overall, the control or activity frameworks that are connected to it. In one context, developers focused on an assault on a sword plant in Germany in December 2014, using boobytrapped messages to remove login accreditations which empower the interloper to access the plant"s control framework. Attackers obtained entrance into the steel industrial facility"s corporate framework utilizing spear-phishing and social designing which thus prompted the disappointment of control parts and all assembling apparatuses were confronted with the inconvenience. DoS is a kind of penetration in Cyber security wherever the attackers (hackers) endeavor to keep legitimate clients from getting the help. In a DoS attack, the attacker ordinarily sends over-the-top messages asking the organization or worker to verify demands that have invalid addresses brought back.

3.3.3 Cloud Computing and Big Data Analytics

This section discusses the challenges of Cloud Computing and Big Data. Aceto et al. (2020) present the various available services from Cloud Computing to maintain Transparency of infrastructure data privacy and performance monitoring. Infrastructure availability is based on various services (Zhong et al., 2017). Cloud Computing enriches the abilities of manufacturing industries states that establishing communications between different clouds reduces the reliability and efficiency of cloud-based systems and also affects resource allocation, compatibility, and availability. Wang and Xu (2013) present interoperable manufacturing cloud solutions. Chen et al. (2017) implement the smart factory concept to meet the intelligent services of data cleaning, data protection, and real-time data processing, and highlights the challenges incurred in giving services.

O'Donovan et al. (2015) identify the need for developing a backup implementation plan for Big Data. Xu et al. (2018) developed a practical application to analyze data in silos from IoT devices and existing ICT (Information and Communication Technologies) that require a robust Big Data platform (Bajic et al., 2020); deficiency in information system standards and system of rules; administration challenges and financial feasibility regarding company size (Lenz et al., 2018); and unrelated data placed into limited storage and integration issues of the database.

3.4 Impact of Industry 4.0 Applications and Projects

This section summarizes the various Industry 4.0 applications as well as some of the good projects from the German Federal Ministry of Education and Research, France, Asia, and European countries.

3.4.1 Impact of Industry 4.0 Applications

This section summarizes the various author works presented in various venues that aspire to improve the quality of human life through various Industrial Revolution applications. The efficient steps followed in Industry 4.0 revolution remove the hindrances that occurred in connectivity issues which may affect the objective of the whole process. Turner et al. (2020) present an application of IR 4.0 on the construction site. Sensor wearable devices are used to predict the risk and alert the workers. Digital Twin Technology is used to assess the structure designed for smart environment conditions. As digital technologies increase, the standards are needed to be improved in the perception of safety, security, and cost. The nature of the devices used in IR4.0 is compact, assured better communication technology, and advanced computational power (Bousdekis et al., 2020). An intelligent system comprising the human, machine, cyber and physical components is presented. The objective of the report is to ease the interactions between the man and machine (AI) and physical and digital worlds in the background of CPSs. The presented framework of HCPS (Human Cyber-Physical Systems) has applied smart technologies in its elements of computer-based intelligence (Artificial Intelligence) in Industries, Digital Twins in CPS, and Operator 4.0-AI Symbiosis. Information and Data-Driven calculations are utilized for Knowledge Acquisition and Data Ingestion for Collaboration between Digital Twin-driven and Human–Machine systems. The ensured benefits through the presented framework are transparency, security, robustness, and controllability. The system had mentioned human intelligence and machine intelligence to have a good HCPS. Afrizal et al. (2020) present an analysis of the component CPS for the industry. The article stresses the importance of integrating cyber-physical and cyber humans. Moreover, the authors introduced 5C Component levels (Connections, Conversation, Cyber, Cognition, and Configuration). The feedback information from the cyber human system is exploited for self-adaptation, corrective, and preventive actions.

Sinha and Roy (2020) present a smart factory concept that integrates the manufacturing sector with IT sectors. The presented framework intends to complement factory automation, thereby facilitating economic growth. The article showcases the various issues in industries and the possible solutions for enhancing the idea of Smart Factory along with controversies faced by CPS in IR4.0 (Jazdi, 2014). Through smart factories, the following benefits are tried to attain: flexibility, energy and resource proficiency, correctness, consistency, security, user-friendliness, and scalability (Saniuk et al., 2021). Enterprises produce flexible products at minimal costs by deploying intelligent technologies such as Big Data, IoT, and AI. Creating greater benefits for the client-side triggers challenges for the industry. An efficient production network called Cyber Industry Network is presented through mobile knowledges and the IoT. Through the Cyber Industry Networks, there is the possibility of increasing the manufacturing of the product using geographically dispersed corporate resources. Pivoto et al. (2021) present a survey of CPS architecture models in the perception of the IIoT. As the complexity of IR 4.0 applications increases, the challenges in ensuring security measures increase as well. Moreover, the paper discusses the advanced methods to build CPS and how the technologies are used to ensure the IR 4.0 attributes through vertical and horizontal integrations. The paper not only highlights issues in CPS building but also enlightens the necessary changes that would take place shortly (Kannengiesser & Müller, 2018). Implementing CPS for the IIoT attains great improvement in processes, products, and value generation.

The advances in digital technologies create an ample number of opportunities for intelligent Industries. There is a lack of protocols for the industries that aim to implement intelligent procedures. Da Silva et al. (2020) summarize the possible scientific procedures for the industries from different venues. In addition to the review of various intelligent procedures, the authors highlight the hindrances reported in different industries. The paper reports that the industries are facing issues due to a lack of enough financial support, infrastructure shortage, and domain expert's' guidance (Khan et al., 2020). The trending technologies, as well as the applications of the IoT, lay the path for the IIoT. The IIoT creates a new vision for IoT in the manufacturing industries. Newly enabled technologies like Blockchain and Cloud Computing are deployed to ensure Robustness, Flexibility, and Trust faced by Industries. To enlighten the readers about the IIoT, a precise definition is presented. The authors described the benchmarking research efforts carried out in IIoT. Eventually, the enabling intelligent technologies for the IIoT and its inherent challenges faced are also summarized.

Industries wish to transform their nature of business in line with PSS (Product-Service System). Gaiardelli et al. (2021) identify the main features that would enrich the shape of PSS IR4.0. There are four measurements in PSS which lead to the advancement of Industry 4.0: (1) value offerings manifestation, (2) customer value experience, (3) value creation mechanism, (4) and value creation interactions. The main idea here is to research the further turns of events and conversation about the PSS and adoption of Industry 4.0 at the two levels, i.e., theory and practical. The merit of such integration brings the new technologies into account and received feedback from the experts (Moeuf et al., 2020). The author discusses the critical success factors and the associated risks and opportunities of Industry 4.0. To achieve objectives like factors, risks, and opportunities of Industry 4.0 in SMEs (small- and medium-sized enterprises), the authors introduced the Delphi–Régnier study. Delphi"s study stated the expert's opinion on specific topics. Régnier"s study had stated a tool that uses a color panel like a traffic signal based on expert's' opinions; 80% of experts had agreed on the technique in SMEs of Industry 4.0.

The IR4.0 and IIoT pave the way to achieve greater improvement for today's industries. The interconnected devices used by the manufacturing sectors suffered from security vulnerabilities due to improper device configuration in the aspect of security. An online tool Shodan is used (Fernández-Caramés & Fraga-Lamas, 2020) as a practical teaching method for Industry 4.0 and IIoT cyber security. Detecting and preventing the Shodan attacks on IIoT devices creates a good learning platform for engineering graduates. understand the impact of security violations. The authors instruct the way to learn cyber security and the IIoT through use cases (Aheleroff et al., 2020). In this paper, the authors specified an IoT-enabled smart appliance through digital transformation – a case study of using the smart refrigerator as an IoT-based home appliance SPS framework under Industry 4.0. Smart refrigerators – with the help of communication signals and IoT networks – are activated through Wi-Fi or Bluetooth so that monitoring and controlling the refrigerator cooling and temperature level, Actuators, and Fridge Cycles became easy (Geng, 2017). Such devices are feasible and affordable. But adding the feature to the existing one incurred more time and cost.

Trending technologies like SDN-IoT and Blockchain are integrated to frame the new architecture namely DistB-SDoIndustry for providing better security for IoT devices. The architecture presented by Rahman et al. (2020) is robust enough against the packet-based attacks like DoS, Flooding Attacks, and the time duration of the packet. The presented architecture uses BCT for data validation and verification. The failure rate had been significantly reduced when compared with the traditional model. Thus, the new technology facilitates attaining efficiency, confidentiality, and higher throughput. Serror et al. (2020) present the possible security vulnerabilities anticipated for the IIoT. The presented report focussed on the security of the IIoT value chain considering the devices and industrial data. For the IIoT, the security aspects like confidentiality, non-repudiation, integrity, privacy, and availability are needed to be ensured on communication channels as well as data. Hence, the authors summarize the challenges and requirements of Industry 4.0 applications amidst the security breaches.

Costa et al. (2020) present the FASTEN IIoT Platform to offer the benefits of adaptable, configurable, and open answers for IIoT applications. FASTEN IIoT offers an intuitive interface for Industry 4.0 applications. In light of AI progressed devices, obtain secure and effectively believed information sharing. The introduced stage includes three pillars: (1) automated robots and manufacturing units, (2) an open IIoT platform, and (3) a real-time environment for predictive modeling.

3.4.2 Smart Projects for Industry 4.0

There have been a few projects running to achieve overall digitization of Industry 4.0.

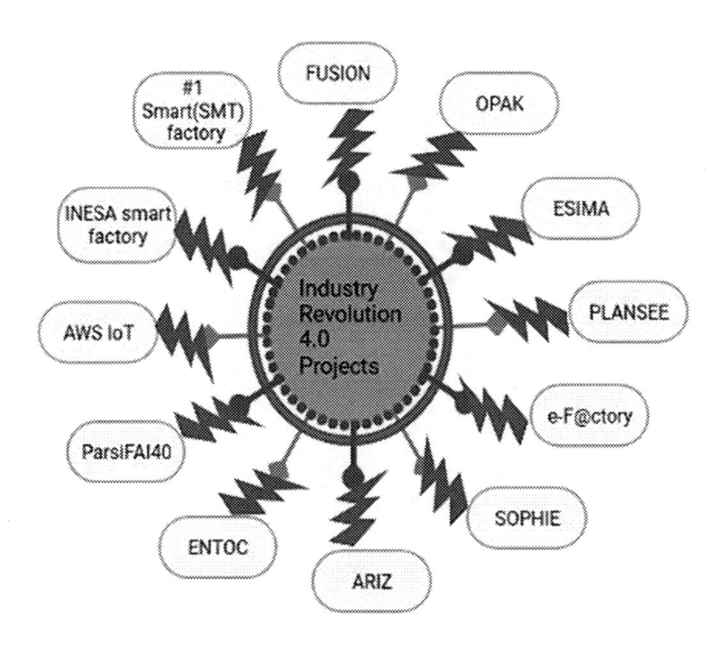

FIGURE 3.5
Industry 4.0 projects.

These projects are sponsored by various government sectors along with industry partnerships. Funding has been done by small- and medium-sized enterprises (SMEs concerned with the particular digital transition). Such projects are primarily concerned with obtaining strategic and organizational efforts for Industry 4.0 implementation (Oztemel & Gursev, 2020). The highlighted smart projects in Industry 4.0 (Figure 3.5) are ENTOC, ARIZ, SOPHIE, e-F@CTORY, PLANSEE, ESIMA, OPAK, FUSION, #1 Smart (SMT) factory, INESA Smart Factory, AWS IoT, and ParisFAI40.

- The project ENTOC's purpose is to make an engineering process so that the result is a simulated ordering which is simpler and more unbroken. This project mainly provides the required tools for simulation and description of components in automation in Industry 4.0.
- ARIZ is also called '"Future Industry Work"' and focuses on the safe human–robot machine, which is used mainly in production workplaces in industries.
- SOPHIE is mainly used in real-time applications of combining the genuine production and digital factory.

- The main aim of E-F@actory is to get a flexible connection among the product manufacturers in the world. This project helps the world in faster communication and increases the robustness of the technologies.
- PLANSEE is one of the Industry 4.0 projects to develop the factory autonomously.
- ESIMA project is used to optimize the efficiency of resources in production systems.
- OPAK Project aims at simple manufacturing processes and mechanisms with digital memory.
- FUSION is meant to encourage collaborative activities between Asian and EU countries.
- #1 Smart (SMT) factor Project will permit the manufacturers to deliver smarter, more adaptable gadgets in a future hardware organization.
- INESA smart factory's main goal is to build smart cities with smart manufacturing companies. This project mainly uses Big Data and the IoT to have error-free problems in manufacturing equipment.
- AWS IoT is used to route the messages reliably and securely in between the AWS endpoints and to different gadgets.
- The ParsiFAl 4.0 research project is meant to develop a thin electronic system labeled S3. The system is built upon micro-controllers, sensors, and compact displays. Such a device can facilitate logistics sectors to find a reliable route for the transport of sensitive goods.

3.5 Application of BCT to Industry 4.0

BCT gained much attention in the finance and payment sectors. Though a lot of questions are remaining unanswered regarding scalability, sustainability, and security, BCT has become one of the key elements for ensuring Internet security in Industries (Parizi & Dehghantanha, 2018). Initially, Blockchain was planned for cryptocurrency and payment systems like Bitcoin, and after that, Blockchain was used for the automation of digital finance through smart contracts (Efanov & Roschin, 2018). And now, Blockchain is a promising solution for complementing the processes meant for Industry 4.0 automation.

Blockchain acts as a database of chunks where these blocks are associated with the hash function. The blocks in a Blockchain are immutable. Every node contains a ledger copy in the network without failure (Alkhalifah et al., 2020). All data are encrypted with cryptographic techniques and add the history chain as a new block. The transactions in the blockchain cannot be changed when they are published in the network. In 2009, the idea of electronic cash was introduced, i.e., Bitcoin. Bitcoin is the cryptocurrency of BCT which allows digital cash to be transferred within the distributed ledger. Other than bitcoin, there are many other cryptocurrencies like XRP (Ripple), ETH (Ethereum), BNB (Binance Coin), LTC (Litecoin), and BCH (Bitcoin Cash). Various blockchain platforms are discussed in Table 3.2.

Innumerable consensus mechanisms are used to add the chain after validating the block with other participants. Consensus mechanisms are

TABLE 3.2

Blockchain Types and the Available Platforms

	Public/Permissionless	**Private/Permissioned**	**Consortium/ Federated**
Decentralization	Completely decentralized	Less decentralized	Less decentralized
Identification of partaker	Yes	No	No
Partaker activity restriction	No limits	Consensus between the Partaker	Yes, based on consensus
Who can do transaction validation	Every Partaker	Only authorized Partaker	Only authorized Partaker
Smart contracts	Yes, depends on the platform	Yes, depends on the platform	Yes, depends on the platform
Accessibility	Anyone	Only one organization	Multiple organizations
Transaction per second	Low	Lighter and faster	Lighter and faster
Consensus protocols	PoW (Proof-of-Work), PoET (Proof-of-Elapsed-Time), and PoS (Proof-of-Stake)	PoS (Proof-of-stake), RAFT, PBFT (Practical Byzantine Fault Tolerance), BFT (Byzantine Fault Tolerance)	Multi-part consensus algorithms or voting schema to reach an agreement
Platforms	Bitcoin, Litcoin, Ethereum, Hyperledger Fabric	Tendermint, Quorum, Hyperledger Fabric, Corda, HydraChain, Kadena, Exonum, Swirld, Symbiont	Ripple, R3 (banks), and B3i (insurance)
Use cases	Healthcare industry, supply chain industry, agriculture industry, manufacturing industry	Healthcare industry, supply chain industry, power industry, manufacturing industry, drone (UAV) industry	Education industry and power industry

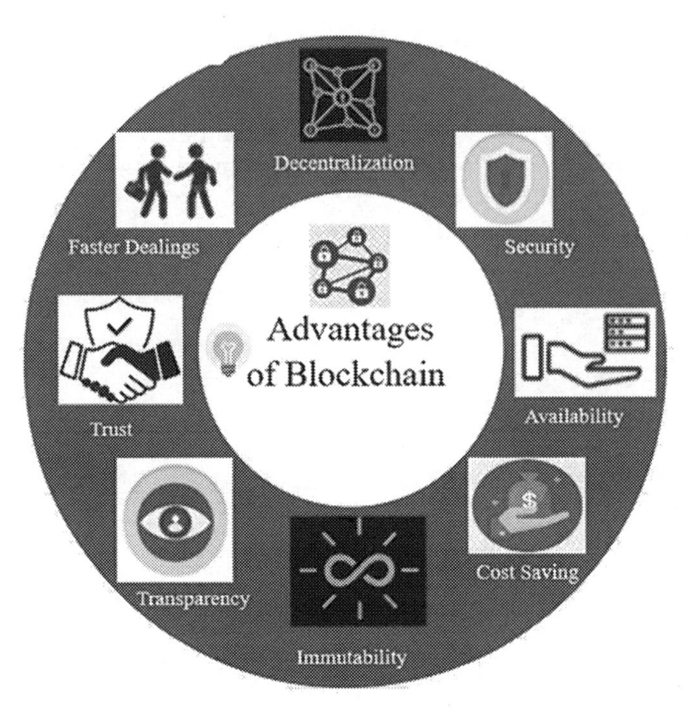

FIGURE 3.6
Advantages of BCT.

intended to allow mistrusting users in a blockchain network to cooperate. This prevents malicious actions without requiring a central authority. Some of the consensus mechanisms include PoW (Proof-of-work), PoET (Proof-of-elapsed-time), and PoS (Proof-of-stake). Some advantages of blockchain are shown in Figure 3.6.

Alkhalifah et al. (2020) state that the victim of blockchain security breaches varies from an individual to banking, health care, and manufacturing sectors. According to the authors, blockchain vulnerabilities are of three types – hack, scam, and smart contract flaws – and the number of events of each type and its associated loss value in terms of millions can be found in Figure 3.7. The vectors of hack are of various types such as email accounts, cloud services, phishing attacks, malicious attacks, and SIM-based attacks. These were all detected and blocked from 2014 to 2018. The privacy of identity property provided by blockchain has allowed itself to become the platform of choice for scams. Figure 3.7 shows that the second-highest loss is due to scams such as the Ponzi Scheme and Pyramid Scheme, causing a loss of 1 billion dollars from 2014 to 2018. The smart contract flaws came into the picture in the years 2014–2018.

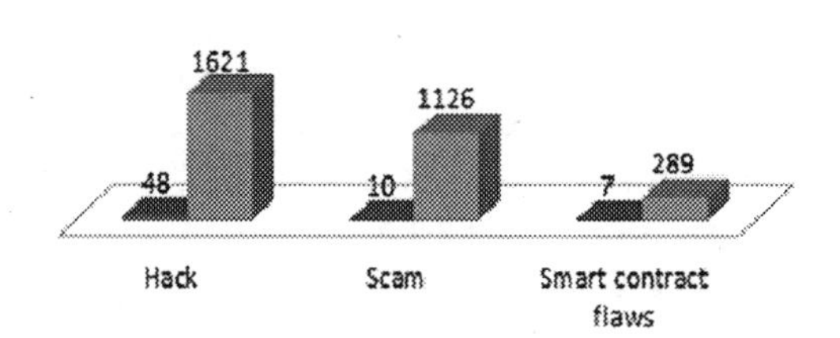

FIGURE 3.7
Blockchain vulnerabilities event and its impact on financial loss.

Hackers can make some adversarial changes in the smart contract scripts to cause financial loss.

3.5.1 Blockchain Frameworks

This section portrays some of the up-to-the-minute blockchain frameworks designed for various Industry 4.0 applications. The efficacy of the designed blockchain framework is assessed based on the level of degree in ensuring the various security aspects of confidentiality, integrity, and availability (Alladi et al., 2019). Blockchain is an emerging technology to protect data and the authors narrate the application BCT imposes many challenges while implementing it. The authors discuss the security aspects like the privacy of participated organizations, scalability of block data, adoption, and integration cost (Xie et al., 2019). Smart environments suffer from the effects of security breaches. The security aspects are improvised using public key infrastructure (PKI). BCT acts as a booster dosage in ensuring security for smart applications like smart citizen, supply chain management, smart grid, smart healthcare, and smart transportation. Authorized entities are highly correctly verified through the blockchain (Yang et al., 2019). Billions of devices are connected to run the applications in an edge computing network. By integrating BCT, network security aspects and computation efficiency are greatly improvised. Distributed databases (BigchainDB, Enigma, and IPFS) are used to solve the storage problem.

Adaption of blockchain for the gas and oil industry may raise transparency and reduce transaction costs. Implementing BCT (Lu et al., 2019) involves a lot of challenges (legal and cyber risks). Such technologies are in an experimental stage (Fraga-Lamas & Fernández-Caramés, 2019). A blockchain framework is presented for the automotive industry and facilitates solving financial-related issues using a smart contract. A consortium blockchain is preferable to communicate with the different organizations with restricted access privileges (Shen & Pena-Mora, 2018). In the smart environment, the likelihood of interconnected devices is a key component for automating the process, where secure data and information exchange play a vital role. Bringing a BCT-enabled network enhances network security (Salman et al., 2018). Distributed network services and challenges are solved through the blockchain framework. Though the amount of network traffic data is huge, along with cloud usage, BCT provides a solution to implement security aspects such as confidentiality and integrity (Ferrag et al., 2018). The IoT can be fine-tuned with different specializations like the IoC (Internet of Cloud), the IoV (Internet of Vehicles), and the IoE (Internet of Energy). Secure communication remains essential for the IoT environment and data privacy is greatly improvised by Fernández-Caramés and Fraga-Lamas (2018), and Christidis and Devetsikiotis (2016) demonstrate the applicability of BCT for the design of smart applications intended for industries (Figure 3.8).

Table 3.3 summarizes the various security frameworks with BCT to cater to the need of many smart Industry 4.0 applications. This discusses bring forth the various blockchain-enabled industry applications along with the ensured security aspects and merits and demerits of the presented framework wherever possible. Figure 3.9 presents the usage of BCT in various sectors.

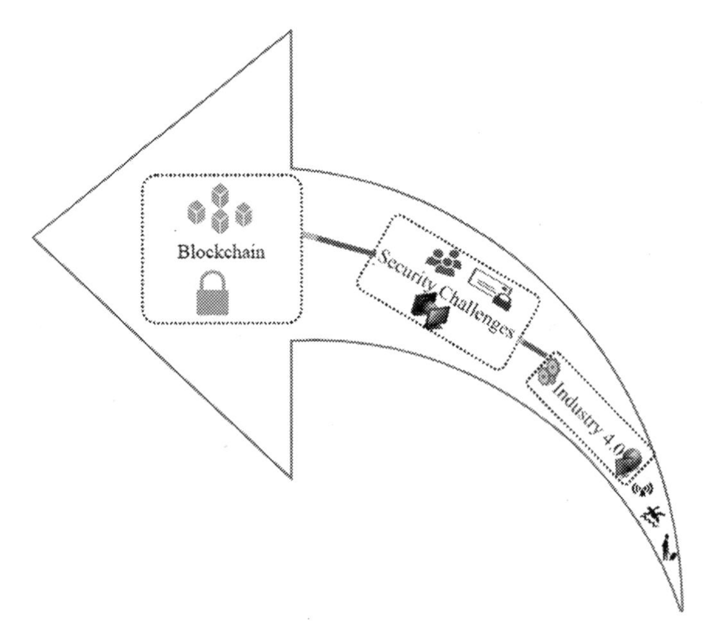

FIGURE 3.8
Application of BCT to various Industry 4.0 applications.

TABLE 3.3

Security Framework for Blockchain Technology

Author Name	Method Used	Ensured Security Aspects	Merits	Demerits
Stodt and Reich (2020)	Presents DCMB (Data Communication Module for Blockchain) for the application of Industry 4.0	Data Confidentiality, Authenticity, and Integrity	No centralized data storage. Smart contracts are enriched with data signatures. Malicious intrusion is prevented.	Availability is an issue not been addressed.
Singh (2020)	Discusses Industry Data Management procedure using BCT.	Confidentiality and Integrity	BCT enabled the database to be robust enough against data security breaches. Discusses the intelligent BCT-enabled database design. The issues associated with Industry data generation and its security issues are highlighted.	Distributed database design challenges are not addressed.
Kumar and Gupta (2020)	BlockEdge framework ensures low latency services along with the security aspects for the IIoT environment.	Confidentiality and Integrity	(BlockEdge) produced good results in terms of performance, resource efficiency, network usage, and power consumption.	The transparency issue is not addressed.
Hang and Kim (2019)	BCT enabled a platform for the IoT environment for ensuring data integrity.	Integrity	Resource-constrained IoT architecture. Easily customized for various IoT use cases. Ensures scalability, high throughput, lightweight, and transparency.	The interoperability issue needs to be addressed. The presented framework is not exploited with another consensus mechanism except proof of concept.

(Continued)

TABLE 3.3 (*Continued*)

Security Framework for Blockchain Technology

Author Name	Method Used	Ensured Security Aspects	Merits	Demerits
Viriyasitavat et al. (2018)	An automated BPM (Business Process Management) solution is provided to regulate the business workflow.	Privacy and Confidentiality	BCT-enabled BPM solution ensured reliability and cost efficiency. Middlemen intervention is not required. Due to automation, the operational cost is reduced.	Not tested with other existing BPM solutions.
Puri et al. (2020)	BCT-based network security framework for IIoT applications in Industry 4.0	Trust and Privacy	Security aspects are improvised due to the presented three-layer architecture.	The applicability of other key components of Industry 4.0 is not disclosed.
Rahman et al. (2020)	DistB-SDoIndustry framework equipped with robust technologies SDN-IoT to enrich the security measures of Industry 4.0 services.	Privacy and Confidentiality	DistB-SDoIndustry offers flexibility. Robust enough against the attack amid high network traffic. The node failure rate is greatly reduced. Obtains high throughput.	Steps to prevent DoS and Flooding Attacks are not yet explored.
Mohamed and Al-Jaroodi (2019)	BCT-enabled Man4Ware is discussed for smart manufacturing.	Integrity and Privacy	Offers flexible and extensible environment above ensuring security aspects.	Though most of the challenging aspects of Industry 4.0 are highlighted, the solutions to address them are few.

Putz et al. (2021)	EtherTwinDApp is presented to remove the challenges incurred in sharing Digital Twins among multiple parties.	Integrity and Confidentiality	Data sharing is greatly improvised without a trusted third party. Suitability is evaluated through	Applicability of the proposal to another domain DT is not explored.
Lee et al. (2019)	Impacts of BCT CPPS Architecture for Industry 4.0 are presented. A three-layer BC architecture for the benefit of Industry 4.0 is introduced.	Confidentiality and Privacy	Elimination of single points of failure. Safe and reliable operations are ensured for Industry 4.0 applications.	Cost issues in storage capacity and implementation difficulty in real-time are not discussed.
Lin et al. (2018)	Belen's fine-grained authentication framework is presented.	Confidentiality and Privacy	The proposed framework is robust against MITM (Man in the Middle), DDoS (Distributed Denial of Service), and Modification and Replay attacks.	The applicability of the presented BSeln for smart factories could be explored.

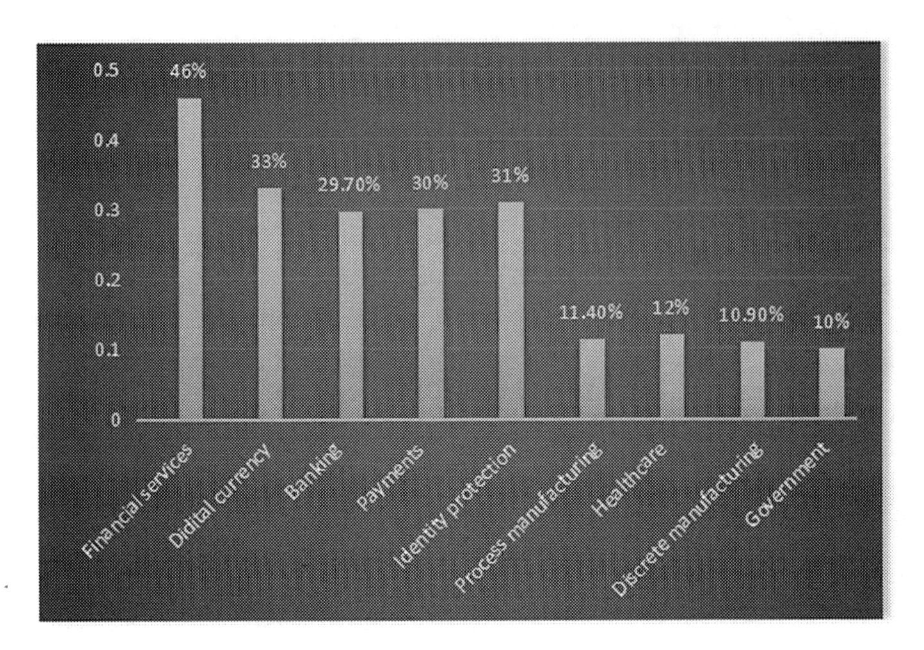

FIGURE 3.9
Usage of blockchain technology in various sectors.

3.5.2 Threats from Blockchain Components

In a blockchain, smart contracts are used to have a contract between the users in an effective manner. For security purposes, we use smart contracts in blockchain but in that also, authors da Silva et al. (2020) highlight the following loopholes:

- Smart contracts are immutable and also, we cannot change the software flaws if any problem occurs.
- By having a signature on a smart contract, the agreement is visible to all the users in the blockchain, and everyone is aware of problems before following on it. Malicious agents will exploit the problems very easily and the codes could be easily accessible.
- Smart contracts work poorly in transactions/transfers of high rates because of exploited bugs.
- Re-entrancy vulnerability is present in smart contracts, as is the case of the DAO incident in 2016. Re-entrancy is a kind of vulnerability displayed in the Ethereum Smart Contract. As the name suggests, an attacker first deposits an amount X to a multiparty smart contract. The attacker then executes a function to withdraw an amount

Y, which is more than X, before the balance of funds deposited and withdrawn has been settled. The effect is the attacker essentially stealing money from other parties in the contract.

Cyber attacks are prevented due to the merits of BCT such as shared ledger, immutability, privacy, consensus, and cryptographic mechanisms, which are provided to prevent different attacks and frauds. Though BCT ensures the tamper-proof wall, hackers still find gaps in its building blocks and damage the assets. Figure 3.10 shows vulnerabilities in BCT components (Adewole et al., 2020).

Some of the problems are as follows:

- Wallet cryptocurrencies are prone to cyber-attack. The hackers can hack easily the clientside wallet. One such example of a wallet hacker is Trojan Devil Robber, which collects information by closing the wallet records.

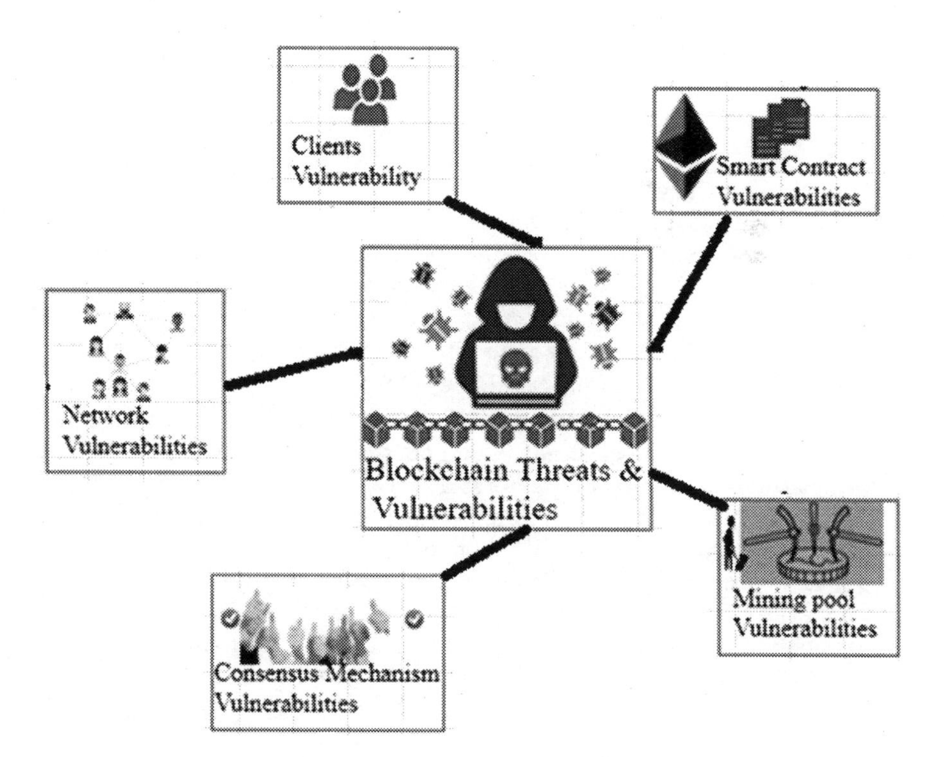

FIGURE 3.10
Types of vulnerabilities in blockchain technology components.

- Double spending is the other possible security threat from BC. According to privacy-centric digital currency Verge (VxG), there are three cases of 51% attacks from April to May 2018, and the attacks usually occur at the target nodes in changing the timestamp. Malicious miners can mine time-stamped blocks, misleading the network to believe that new blocks were mined and embedding them into the blockchain. The following mined block is additionally, quickly added to the network. The attacker had the option to mine one block each second thusly and gathered 250,000 VxG in the prime attack on April 4, 2018.
- 51% attack is the other possible attack on the blockchain network. To execute such an attack, high computation power is essential. Such an attack may occur when the miner has controlling power of more than 50% of the hashing power of a network and is able to generate a network fork and complete an attack of double-spend. The mining attacker will erase the transaction history and prevent the confirmation of the new block.
- In a double-spending attack, the attacker uses the same currency for multiple transactions. The transactions are reversed in this attack and leads to spending the money again. It mainly prevents the miners from mining the valid blocks. In an example of this attack in 2018, Bitcoin Gold lost $17.5 million.
- In blockchain networks, the other anticipated attack is Selfish Mining. Such an attack targets blockchain network integrity. In this context, the sapper effectively mines a chunk but doesn't reveal it to the additional chunks. The sapper adds such blocks in his private chain to make it lengthy. Compared to the smallest chain, the longest chain is only considered valid. In Japan, such an attack was attempted in 2018. Through this attack, a person with a hidden chain could get profit.
- During the software update, some of the nodes in the transaction view older versions whereas some other nodes receive the newer one; such an issue is known as a fork. Forks are provided by miners or cryptocurrency users. The fork is split. The fork is used to create a new network. Cryptocurrencies split their coins to produce new coins. Let's take an example where Bitcoin went through a fork on 24th October 2017 to generate a novel cryptocurrency called Bitcoin Gold. The intention is to facilitate many people to work on a different version. But it's a debatable topic as to whether the fork is good or bad.

The disadvantages of cryptocurrencies and Blockchains are as follows:

- Volatile Market: Based on the trend, the cryptocurrencies' values are subject to change within a few hours; to prevent the loss, make transactions when their value increases.
- Poor Security: Loss of wallet money occurs when the security measures are not sufficient. Since cryptocurrency programmers are not good with security aspects, hackers put their knowledge to crack the codes.
- Lack of Solid Anonymity: Cybercriminals may use anonymity for money laundering and human trafficking.
- Prone to Scams: The owner has the right to access their wallet in blockchain with the private key. In case it is lost or forgotten, even the holder can't open it. According to reports, from 2011 to 2014, many scams with a total amount of more than 10 million U.S. dollars involved cryptocurrency transactions including high-yield investment schemes such as investments in mining, Ponzi schemes, "duplicitous wallet" payments, and exchange scams.
- Trust As a Savings Point: Elderly people are reluctant to use cryptocurrencies since most of the guidelines are not well defined and the processes involved in it are not practically feasible.

3.6 Conclusion

The arrival of Industry 4.0 facilitates industries to obtain significant growth in many aspects. Equipping Industry 4.0 applications amid security challenges is possible through BCT 4.0-enabled security solutions. Amid the incompatibility issues in the interconnected components, well-defined protocols and configured devices will pave the way for successful automation. This chapter enlightens the readers about the key components of Industry 4.0 applications and their associated inherent challenges in terms of development and security breaches. Moreover, this chapter stresses the importance of deploying a BCT-enabled framework for Industry 4.0 applications and summarizes the ongoing BCT-enabled smart applications along with their pros and cons wherever possible.

References

Aceto, G., Persico, V., & Pescapé, A. (2020). Industry 4.0 and health: Internet of things, big data, and cloud computing for healthcare 4.0. *Journal of Industrial Information Integration*, 18, 100129.

Adewole, K., Saxena, N., & Bhadauria, S. (2020). Application of Cryptocurrencies Using Blockchain for e-Commerce Online Payment. In *Blockchain for Cybersecurity and Privacy* (pp. 263–305). CRC Press.

Afrizal, A., Mulyanti, B., & Widiaty, I. (2020, April). *Development of Cyber-Physical System (CPS) implementation in Industry 4.0.* In *IOP Conference Series: Materials Science and Engineering* (Vol. 830, No. 4), p. 042090. IOP Publishing

Aheleroff, S., Xu, X., Lu, Y., Aristizabal, M., Velásquez, J. P., Joa, B., & Valencia, Y. (2020). IoT-enabled smart appliances under Industry 4.0: A case study. *Advanced Engineering Informatics*, 43, 101043.

Alkhalifah, A., Ng, A., Kayes, A. S. M., Chowdhury, J., Alazab, M., & Watters, P. A. (2020). A Taxonomy of Blockchain Threats and Vulnerabilities. In *Blockchain for Cybersecurity and Privacy* (pp. 3–28). CRC Press.

Alladi, T., Chamola, V., Parizi, R. M., & Choo, K. K. R. (2019). Blockchain applications for Industry 4.0 and industrial IoT: A review. *IEEE Access*, 7, 176935–176951.

Anisetti, M., Ardagna, C., Cremonini, M., Damiani, E., Sessa, J., & Costa, L. (2020). *Security Threat Landscape.*

Bajic, B., Rikalovic, A., Suzic, N., & Piuri, V. (2020). Industry 4.0 Implementation challenges and opportunities: A managerial perspective. *IEEE Systems Journal*, 15(1), 546–559.

Bousdekis, A., Apostolou, D., & Mentzas, G. (2020). A human cyber physical system framework for operator 4.0–artificial intelligence symbiosis. *Manufacturing Letters*, 25, 10–15.

Carvalho, A. C. P., Carvalho, A. P. P., & Carvalho, N. G. P. (2020). Industry 4.0 Technologies: What Is Your Potential for Environmental Management?. In *Industry 4.0: Current Status and Future Trends* (pp. 29).

Chen, B., Wan, J., Shu, L., Li, P., Mukherjee, M., & Yin, B. (2017). Smart factory of Industry 4.0: Key technologies, application case, and challenges. *IEEE Access*, 6, 6505–6519

Chiu, Y. C., Cheng, F. T., & Huang, H. C. (2017). Developing a factory-wide intelligent predictive maintenance system based on Industry 4.0. *Journal of the Chinese Institute of Engineers*, 40(7), 562–571.

Christidis, K., & Devetsikiotis, M. (2016). Blockchains and smart contracts for the internet of things, *IEEE Access*, 4, 2292–2303.

Costa, F. S., Nassar, S. M., Gusmeroli, S., Schultz, R., Conceição, A. G., Xavier, M., … & Dantas, M. A. (2020). FASTEN IIoT: An open real-time platform for vertical, horizontal and end-to-end integration, *Sensors*, 20(19), 5499.

da Silva, T. B., de Morais, E. S., de Almeida, L. F. F., da Rosa Righi, R., & Alberti, A. M. (2020a). Blockchain and Industry 4.0: Overview, Convergence, and Analysis. In *Blockchain Technology for Industry 4.0* (pp. 27–58).

da Silva, V. L., Kovaleski, J. L., Pagani, R. N., Silva, J. D. M., & Corsi, A. (2020b). Implementation of Industry 4.0 concept in companies: Empirical evidences. *International Journal of Computer Integrated Manufacturing*, 33(4), 325–342.

Efanov, D., & Roschin, P. (2018). The all-pervasiveness of the blockchain technology, *Procedia Computer Science*, 123, 116–121. Available: http://www.sciencedirect.com/science/article/pii/S1877050918300206, doi: 10.1016/j.procs. 2018.01.019

Erboz, G. (2017). How to Define Industry 4.0: Main Pillars of Industry 4.0. In *Managerial Trends in the Development of Enterprises in Globalization era* (pp. 761–767).

Fernández-Caramés, T. M., & Fraga-Lamas, P. (2018). A review on the use of blockchain for the internet of things. *IEEE Access*, 6, 32979–33001.

Fernández-Caramés, T. M., & Fraga-Lamas, P. (2020). Use case based blended teaching of IIoT cybersecurity in the Industry 4.0 era. *Applied Sciences*, 10(16), 5607.

Ferrag, M. A., Derdour, M., Mukherjee, M., Derhab, A., Maglaras, L., & Janicke, H. (2018). Blockchain technologies for the internet of things: Research issues and challenges. *IEEE Internet of Things Journal*, 6(2), 2188–2204.

Fraga-Lamas, P., & Fernández-Caramés, T. M. (2019). A review on blockchain technologies for an advanced and cyber-resilient automotive industry. *IEEE Access*, 7, 17578–17598.

Gaiardelli, P., Pezzotta, G., Rondini, A., Romero, D., Jarrahi, F., Bertoni, M., ... & Cavalieri, S. (2021). Product-service systems evolution in the era of Industry 4.0. *Service Business*, 15(1), 177–207.

Geng, H. (2017). *Internet of Things (IOT)-Based Cyber–Physical Frameworks for Advanced Manufacturing and Medicine*, Wiley Telecom, 545–561.

Georgios, L., Kerstin, S., & Theofylaktos, A. (2019). Internet of things in the context of Industry 4.0: An overview. *International Journal of Entrepreneurial Knowledge*, 7(1), 4–19.

Hang, L., & Kim, D. H. (2019). Design and implementation of an integrated IoT blockchain platform for sensing data integrity. *Sensors*, 19(10), 2228.

Jazdi, N. (2014, May). *Cyber physical systems in the context of Industry 4.0*. In *2014 IEEE International Conference on Automation, Quality and Testing, Robotics* (pp. 1–4). IEEE.

Kannengiesser, U., & Müller, H. (2018, August). *Multi-level, viewpoint-oriented engineering of cyber-physical production systems: An approach based on Industry 4.0, system architecture and semantic web standards*. In *2018 44th Euromicro Conference on Software Engineering and Advanced Applications (SEAA)* (pp. 331–334). IEEE.

Keerthi, C. K., Jabbar, M. A., & Seetharamulu, B. (2017, December). *Cyber physical systems (CPS): Security issues, challenges and solutions*. In *2017 IEEE International Conference on Computational Intelligence and Computing Research (ICCIC)* (pp. 1–4). IEEE.

Khan, W. Z., Rehman, M. H., Zangoti, H. M., Afzal, M. K., Armi, N., & Salah, K. (2020). Industrial internet of things: Recent advances, enabling technologies and open challenges. *Computers & Electrical Engineering*, 81, 106522.

Kumar, A., & Gupta, D. (2020). Challenges within the Industry 4.0 Setup. In *A Roadmap to Industry 4.0: Smart Production, Sharp Business and Sustainable Development* (pp. 187–205). Springer, Cham.

Kumar, T., Harjula, E., Ejaz, M., Manzoor, A., Porambage, P., Ahmad, I., ... & Ylianttila, M. (2020). BlockEdge: Blockchain-edge framework for industrial IoT networks. *IEEE Access*, 8, 154166–154185.

Lee, J., Azamfar, M., & Singh, J. (2019). A blockchain enabled cyber-physical system architecture for Industry 4.0 manufacturing systems. *Manufacturing Letters*, 20, 34–39.

Lenz, J., Wuest, T., & Westkämper, E. (2018). Holistic approach to machine tool data analytics. *Journal of Manufacturing Systems*, 48, 180–191.

Lin, C., He, D., Huang, X., Choo, K. K. R., & Vasilakos, A. V. (2018). BSeIn: A blockchain-based secure mutual authentication with fine-grained access control system for Industry 4.0. *Journal of Network and Computer Applications*, 116, 42–52.

Lu, H., Huang, K., Azimi, M., & Guo, L. (2019). Blockchain technology in the oil and gas industry: A review of applications, opportunities, challenges, and risks. *IEEE Access*, 7, 41426–41444.

Moeuf, A., Lamouri, S., Pellerin, R., Tamayo-Giraldo, S., Tobon-Valencia, E., & Eburdy, R. (2020). Identification of critical success factors, risks and opportunities of Industry 4.0 in SMEs. *International Journal of Production Research*, 58(5), 1384

Mohamed, N., & Al-Jaroodi, J. (2019, January). *Applying blockchain in Industry 4.0 applications*. In *2019 IEEE 9th Annual Computing and Communication Workshop and Conference (CCWC)* (pp. 0852–0858). IEEE.

O'Donovan, P., Leahy, K., Bruton, K., & O'Sullivan, D. T. (2015). An industrial big data pipeline for data-driven analytics maintenance applications in large-scale smart manufacturing facilities. *Journal of Big Data*, 2(1), 1–26.

Oztemel, E., & Gursev, S. (2020). Literature review of Industry 4.0 and related technologies. *Journal of Intelligent Manufacturing*, 31(1), 127–182.

Parizi, R. M., & Dehghantanha, A. (2018). *Smart contract programming languages on blockchains: An empirical evaluation of usability and security*, In *Proc. Int. Conf. Blockchain (ICBC), in Lecture Notes in Computer Science*, vol. 10974, Chen, S., Wang, H., & Zhang, L. J., Eds. Cham, Switzerland: Springer (pp. 75–91).

Parizi, R. M., Dehghantanha, A., Choo, K.-K. R., & Singh, A. (2018). *Empirical vulnerability analysis of automated smart contracts security testing on blockchains*. In *Proc. 28th Annu. Int. Conf. Comput. Sci. Softw. Eng. (CASCON)* (pp. 103–113).

Pivoto, D. G., de Almeida, L. F., da Rosa Righi, R., Rodrigues, J. J. P. C., Lugli, A. B., & Alberti, A. M. (2021). Cyber-physical systems architectures for industrial internet of things applications in Industry 4.0. *Journal of Manufacturing Systems*, 58, 176–192.

Puri, V., Priyadarshini, I., Kumar, R., & Kim, L. C. (2020, March). *Blockchain meets IIoT: An architecture for privacy preservation and security in IIoT*. In *2020 International Conference on Computer Science, Engineering and Applications (ICCSEA)* (pp. 1–7). IEEE.

Putz, B., Dietz, M., Empl, P., & Pernul, G. (2021). Ethertwin: Blockchain-based secure digital twin information management. *Information Processing & Management*, 58(1), 102425.

Rahman, A., Sara, U., Kundu, D., Islam, S., Islam, M., Hasan, M., ... & Nasir, M. K. (2020). Distb-SdoIndustry: Enhancing security in Industry 4.0 services based on distributed blockchain through software defined networking-iot enabled architecture. arXiv preprint arXiv:2012.10.

Salman, T., Zolanvari, M., Erbad, A., Jain, R., & Samaka, M. (2018). Security services using blockchains: A state of the art survey. *IEEE Communications Surveys & Tutorials*, 21(1), 858–880.

Saniuk, S., Saniuk, A., & Cagáňová, D. (2021). Cyber industry networks as an environment of the Industry 4.0 implementation. *Wireless Networks*, 27(3), 1649–1655.

Serror, M., Hack, S., Henze, M., Schuba, M., & Wehrle, K. (2020). Challenges and opportunities in securing the industrial internet of things. *IEEE Transactions on Industrial Informatics*, 17(5), 2985–2996

Shen, C., & Pena-Mora, F. (2018). Blockchain for cities—A systematic literature review. *IEEE Access*, 6, 76787–76819.

Singh, M. (2020). Blockchain Technology for Data Management in Industry 4.0. In *Blockchain Technology for Industry 4.0* (pp. 59–72). Springer, Singapore.

Sinha, D., & Roy, R. (2020). Reviewing cyber-physical system as a part of smart factory in Industry 4.0. *IEEE Engineering Management Review*, 48(2), 103–117

Sittón-Candanedo, I., Alonso, R. S., Rodríguez-González, S., Coria, J. A. G., & De La Prieta, F. (2019, May). *Edge computing architectures in Industry 4.0: A general survey and comparison*. In *International Workshop on Soft Computing Models in Industrial and Environmental Applications* (pp. 121–131). Springer, Cham.

Stodt, J., & Reich, C. (2020). Data confidentiality in P2P communication and smart contracts of blockchain in Industry 4.0. arXiv preprint arXiv:2007.14195.

Tay, S. I., Lee, T. C., Hamid, N. Z. A., & Ahmad, A. N. A. (2018). An overview of Industry 4.0: Definition, components, and government initiatives. *Journal of Advanced Research in Dynamical and Control Systems*, 10(14), 1379–1387.

Turner, C. J., Oyekan, J., Stergioulas, L., & Griffin, D. (2020). Utilizing Industry 4.0 on the construction site: Challenges and opportunities. *IEEE Transactions on Industrial Informatics*, 17(2), 746–756.

Viriyasitavat, W., Da Xu, L., Bi, Z., & Sapsomboon, A. (2018). Blockchain-based business process management (BPM) framework for service composition in Industry 4.0. *Journal of Intelligent Manufacturing*, 31(7), 1737–1748.

Wang, X. V., & Xu, X. W. (2013). An interoperable solution for cloud manufacturing. *Robotics and Computer-Integrated Manufacturing*, 29(4), 232–247.

Xie, J., Tang, H., Huang, T., Yu, F. R., Xie, R., Liu, J., & Liu, Y. (2019). A survey of blockchain technology applied to smart cities: Research issues and challenges. *IEEE Communications Surveys & Tutorials*, 21(3), 2794–2830.

Xu, L. D., Xu, E. L., & Li, L. (2018). Industry 4.0: State of the art and future trends. *International Journal of Production Research*, 56(8), 2941–2962.

Yang, R., Yu, F. R., Si, P., Yang, Z., & Zhang, Y. (2019). Integrated blockchain and edge computing systems: A survey, some research issues and challenges. *IEEE Communications Surveys & Tutorials*, 21(2), 1508–1532.

Zhang, Z., Wang, X., Wang, X., Cui, F., & Cheng, H. (2019). A simulation-based approach for plant layout design and production planning. *Journal of Ambient Intelligence and Humanized Computing*, 10(3), 1217–1230.

Zhong, R. Y., Xu, X., Klotz, E., & Newman, S. T. (2017). Intelligent manufacturing in the context of Industry 4.0: A review. *Engineering*, 3(5), 616–630.

Zhou, K., Liu, T., & Zhou, L. (2015, August). *Industry 4.0: Towards future industrial opportunities and challenges*. In *2015 12th International Conference on Fuzzy Systems and Knowledge Discovery (FSKD)* (pp. 2147–2152). IEEE.

4

The Role of Cutting-Edge Technologies in Industry 4.0

Imdad Ali Shah
Taylor's University, Subang Jaya, Selangor, Malaysia

Noor Zaman Jhanjhi
SoCIT Taylor's University, Selangor, Malaysia

Fathi Amsaad
Eastern Michigan University, Ypsilanti, Michigan, USA

Abdul Razaque
International Information Technology University, Almaty, Kazakhstan

CONTENTS

4.1 Introduction

Since the beginning of the Industrial Revolution, some of the technological advances that have emerged over the past few centuries include steam engines, assembly lines, and computers, all of which aim to produce increasingly powerful technologies and increase productivity and efficiency.

DOI: 10.1201/9781003203087-4

The Fourth Industrial Revolution (Industry 4.0) has changed this paradigm and sparked a revolution by putting less of a focus on technology and believing that the true potential for progress lies in the partnership of humans and robots [1]. Published a paper on the birth of Industry 4.0, stating that Industry 4.0 is a high-tech approach to automated manufacturing using the Internet of Things (IoT) to create smart factories, but has several drawbacks. Therefore, they propose "Industry 4.0" to democratize the co-production of knowledge in big data through a new symmetric innovation model. Their planned Industry 4.0 system leverages the IoT [2]. The concept of Industry 4.0 subsequently gained attention [3]. The authors argue that the urgent need to improve efficiency without displacing human workers from manufacturing poses major problems for the global economy. To address these issues, they came up with the concept of Industry 4.0, which sees human–machine connectivity as a collaboration rather than a competition [4]. Claims that previous discussions of Industry 4.0 did not focus on organizational challenges stemming from human–robot collaboration; at the same time, they discussed potential issues related to human–robot collaboration from the perspective of organizations and employees. Many future studies on organizational robots will undoubtedly be based on their findings [5]. The cutting-edge refers to technological innovations such as logic devices or gadgets. Technological achievements or programs take advantage of the latest and most advanced IT developments. It is at the forefront of leading IT industry innovation [6]. Figure 4.1 gives an overview of the cutting-edge.

We deeply discuss and analyze the following points in this chapter:

1. We discuss the presenting several terms of Industry 4.0 from the literature
2. We analyze the application of Industry 4.0
3. We discuss the cutting-edge technology supporting Industry 4.0
4. We discuss development operations, challenges in terms of security, and human–robot collaboration

FIGURE 4.1
Overview of cutting-edge.

4.2 Literature Review

The term "Society 4.0" refers to a new monetary society based on seeker–gatherer, peaceful farming, and contemporary and data-driven social systems. Society 5.0 is both a development strategy for Japan and a design for the country's future public. It was drafted by Science and Technology. The purpose of Society 5.0 is to create a human society where people can enjoy a truly vibrant and acceptable quality of life while achieving financial goals and addressing societal challenges [7]. It is the broad public that will meet the various needs of individuals regardless of their location, age, gender, dialect, or other factors, by providing key commodities and management. The combination of the internet with current reality (physical space) to produce excellent information and, from there, create new qualities and solve difficult answers is the way to identify it [8]. Its goals are similar to the United Nations Development Programmer's Sustainable Development Goals (SDGs). Although arrangements will vary depending on the financial situation or social basis of a country or region, the SDGs are an overarching goal of progress. Society 5.0 is Japan's way of understanding part of the goals defined by the SDGs [9]. Meanwhile, the sociological issues that Japan is dealing with include an aging population, a dropping birth rate, a shrinking population, and an aging foundation. From this perspective, Japan is one of the first countries to face these challenges [10]. Industry 4.0 investigation reveals a great deal of ambiguity about what it will deliver and how it will disrupt business in detail, but it will break down barriers between the physical and virtual worlds [11]. The next step in the Industrial Revolution, according to Universal Robots' Chief Technology Officer, will be required due to consumers' strong need for individualization in the products they buy. He backed up his claim with a quote from a magazine. According to *Bloomberg*, a German carmaker is already providing additional space for individuals within the vehicle. Customization is a significant factor with current consumers, according to production facilities [12]. Because by 2020, 85 percent of the participants wanted their factories to have a collaborative production line involving people and robots (Atwell, 2017) [13]. Customers will pay the most for products that have a distinct stamp of human care and skill, such as designer items of all kinds, luxury watches, artisan breweries, and black salt from Iceland hand-colored with local coal. Because consumers want to show their personality via the things they buy, the desire for human touch will continue to rise in the future. This describes a new type of personalization, a sense of luxury society, with which businesses must contend [14]. Industry 4.0 is focused on merging human innovation. Every life behavior will be translated with artificial intelligence and then altered with millions of data over the internet in the ERA 5.0 Society [14]. The translation's outcomes will be dedicated to a new wisdom that will strengthen the human potential and expand humanity's prospects [15]. Changes in global economic flows are one of the factors that influence people's behavior [16]. Indonesia's contemporary economic existence has been dissolved in global economic conditions, with growing or falling global economic conditions having an impact on Indonesia's GDP [17]. The term "Society 5.0" refers to changes in people's behavior as a result of technological advancements that are oriented toward human needs. Every configuration of community needs in the ERA 5.0 Society will be human-centered and based on Japanese technology [18,19]. Society 5.0 is a crucial concept in Keidanren's comprehensive action plan to reconstruct Japan. Industry giants including Hitachi, NEC, Fujistu, Toyota, and Panasonic, among others, have incorporated Society 5.0 into their overall objectives [20]. Japan is pursuing Society 5.0 by incorporating digital technologies into a range of platforms and speeding up its implementation to build a society in which all residents can participate in the system [21]. We can clearly understand how strong technology has been influencing our world by looking at where we are now. Scientists have developed cutting-edge technologies over the years to make life easier for people. Technological developments occur at such a rapid pace that people's current lifestyles appear to be out of date in comparison to these advancements. Are you ready to catch a glimpse of what the future may hold (Figure 4.2)?

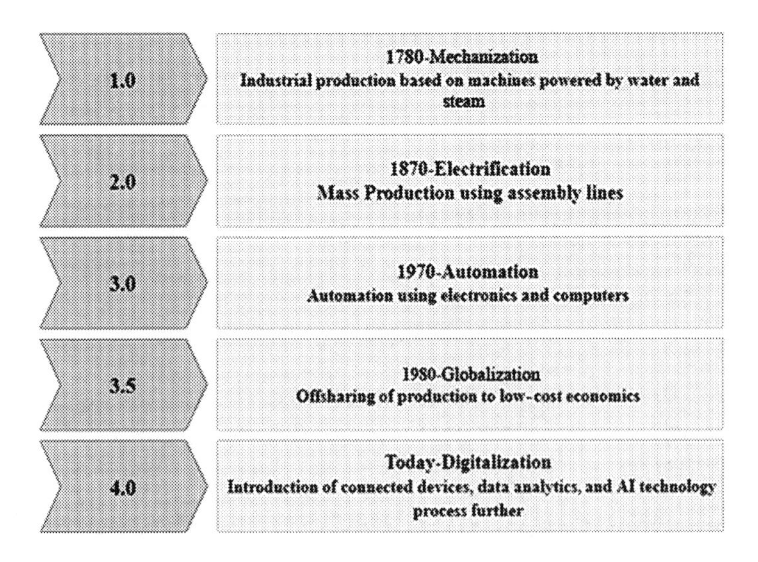

FIGURE 4.2
History of Industrial Revolution from 1.0 to 4.0.

4.3 The Evolution of Industry 4.0

Industry 4.0 is the set of technologies that organizations need to increase their innovation ambitions and respond quickly to market changes. Predictive analytics, interconnect, machine learning, and digital technologies are key areas of focus to transform the way businesses operate and thrive. From suppliers to consumers, Industry 4.0 will enable you to collect, analyze, and use information in real-time with the aim of refining and improving operations, designs, and products through rapid feedback, thereby increasing productivity. Therefore, this chapter advocates and encourages businesses to use Industry 4.0. This chapter examines the current state of business intelligence (BI) technology, how it benefits organizations at the economic and commercial level in terms of decision-making, and highlights some success stories of implementation in a variety of business, government, social, and academic settings. It also looks at industries. The future of Industry 4.0, the field of business intelligence, and how companies should prepare for this shift are discussed. Details on the current state of Industry 4.0 and its benefits, as well as rapid technological breakthroughs, smart digital technologies, and manufacturing integration, will be provided [21]. Industry 4.0 aims to combine high-speed, precise machines with human critical and cognitive thinking. Another important addition to Industry 4.0 is mass personalization, where customers can choose individualized. The goal of human–robot collaboration is to rapidly increase production. Industry 4.0 supports more skilled occupations than Industry 4.0 due to knowledge specialists working with machines. Humans will guide robots in Industry 4.0, which focuses primarily on mass customization [22]. Industry 4.0 already involves robots in mass production, but Industry 4.0 is primarily focused on improving customer satisfaction (Figure 4.3).

FIGURE 4.3
Overview Industry 4.0. [23]

4.4 An Overview of Cutting-Edge Technologies and AI

The rapid development of global technology and its pervasiveness in everyday life has transformed all kinds of things, including online supermarket shopping, car buying and selling, causation technology, access to financing, and more. Customers are one of the most important parts of any company [24]. Innovative technology for the creation of corporate software applications uses the most up-to-date technology. It is the study and design of computer systems capable of carrying out activities that would ordinarily require human intelligence. Visual perception, speech recognition, decision-making, and language translation are among these tasks. Systems capable of doing such tasks are gradually making their way from research labs to businesses. AI is distinct in that it may be applied in a variety of ways [25]. The growing availability of IoT sensors, combined with improved network access, has led to a massively novel digital signal from which enterprises may profit. IoT data in agriculture can help enhance yields and make operations more ecologically friendly. IoT in manufacturing helps assembly line gadgets work more efficiently. Every day, billions of sensors send massive volumes of data to the cloud, from smart homes to smart cities, and industrial sensors to self-driving automobiles [26–30]. Artificial neural networks are a popular machine learning technique. These networks have an input layer, a hidden layer, and an output layer and are inspired by biological neural networks [31–33]. The input layer is responsible for receiving data and signals, while the hidden layer is responsible for data decomposition and processing.

4.4.1 Natural Language Processing

It is a branch of artificial intelligence that aims to bridge the gap between human and computer communication. natural language processing (NLP) is used by AI-enabled systems like IBM's Watson to interpret and respond to human language nuances [34]. This enables more precise data processing and sharing of findings. Figure 4.4 shows an overview of NLPs.

Customer interaction chat-bots	Financial chat-bots	Virtual assistant
Commutation systems	Legal assistants	Cognitive retail
Personal assistants	Web speech	

FIGURE 4.4
Overview of NLP.

4.5 IoT and Cutting-Edge

The IoT, also known as the "new industrial revolution," has expanded its reach to include everything from businesses to homes [35,36]. When it comes to the IoT, however, cloud computing has considerable drawbacks. Advanced 5G technology hasn't been introduced rapidly enough to keep up with the growth of IoT devices. While most big cities have sufficient network connections to enable IoT, rural areas, which house a large quantity of industrial equipment, do not. In the event of a network breakdown, IoT devices in these areas may lose some or all of their functionality, and that's if the area is covered at all. The IoT presents a potential opportunity to build effective production systems and functions using smartphone and sensor monitoring technologies. A broad range of IoT technologies has been developed and deployed in recent years [37–39]. This study critiques current IoT functions in businesses and discusses research breakthroughs and obstacles to recognizing the applicability of IoT in industries. The IoT is expected to emerge as a promising option for dramatically changing the function and operation of existing industrial structures, such as production structures and transportation systems [40]. For example, clever transportation structures can be created when the transportation authority can fine-tune the current location of each vehicle, look outlandishly focused on its movement, and anticipate its future region and likely street traffic.

4.6 Blockchain and Cutting-Edge

According to experts, blockchain and new coins are based on cutting-edge technologies. Such bitcoins can disrupt the global financial system as banking giants make an effort to communicate with their opportunities [41]. It has the potential to lower bank fees while also encouraging greater competition, which charges will be pushed down. Blockchains are mutual, immutable "ledgers" that ensure data integrity and let everyone obtain precise monetary accounting [42]. Blockchain, or so-called distributed ledger technology that underpins transactions in bitcoin and other cryptocurrencies, could facilitate transactions in other assets, posing challenges for banks that charge large fees to help customers exchange currencies and other assets [43]. Payment settlement and clearing are two crucial financial issues products how blockchain has the potential to make a difference. However, a recent Moody study suggests that while blockchain may reduce counterparty risk for financial institutions, it may also increase competition in the banking industry.

4.7 Challenges and Future Implementation

Through a cognitively enabled production process, Industry 4.0 can provide the most personalized services to the client. Some of the future problems highlighted in this section must be solved for services to be seamless [44]. Security is one of the issues that could arise. As we move toward increasingly digitized computing, security vulnerabilities in incompatible data dealing and cloud services for various users and industrial system integration must be cross-checked [45]. Bringing manpower back to the production line may be helpful, customized customer assistance with human–robot co-working must account for the issues associated with scaling up the clients and production methods. Furthermore, the ethical problems surrounding AI adoption should be considered to minimize potential disadvantages and harmful societal consequences.

4.8 Discussion

A similar situation exists for clinical decision-making: Currently, it is impossible to make reliable assessments of complex situations. Real-life patients, however, may face a more challenging set of questions due to the complications of multiple diseases, requiring thorough examination from multiple perspectives [46]. In the field of rehabilitation, AI offers a wide range of potential applications, including not just individuals with a stroke-related disability, but also those who suffer from physical dysfunction due to a variety of factors. When AI systems are coupled with rehabilitative activities, many of them become more viable. It is crucial to note that all disciplines must work together to successfully enhance desired neurological function in different forms of neurological disorders [47–49]. At present, the application of artificial intelligence technology in the field of stroke is still in its infancy, and a lot of basic work needs to be done before more research can be done. Given that emerging technology has demonstrated tremendous learning and generalization capabilities, it may one day gain human-level intelligence [45]. Once AI-based diagnostic techniques and making decision systems are developed, patients will benefit greatly as it will lead to cheaper and higher-quality healthcare. Artificial intelligence technology will not only improve the quality of medical care in primary care facilities but also achieve regional unification of medical care.

Our contribution to this chapter is that we are on the brink of a technological revolution that will revolutionize our way of life and each other's interactions. Cutting-edge technology refers to technical equipment, processes, or accomplishments that make use of the most recent and advanced IT breakthroughs, or in other words, technology at the level of teaching. "Cutting-edge" is a term used to describe leading and innovative IT sector firms. It is commonly acknowledged that both the applied sciences and implementations of the IoT are still in their infancy. Nonetheless, industrial research has several issues, including technology, standardization, security, and privacy. Ensure that IoT devices in industrial settings are compatible. Before IoT, a thorough understanding of industry characteristics and requirements on cost, security, privacy, and risk will be required, and AI will be widely used in many industries. Due to time and efficiency constraints, designing an IoT service-oriented architecture (SOA) is a big task. Additionally, as additional physical artifacts are linked to the network, scaling concerns develop. Scalability is problematic at various distances when there are a large number of problems. Content storage and networking, automation and control, and service provision. From a network perspective, IoT is a network consisting of enormous complexity that consists of connecting more than a few kinds of networks via various spoken communication protocols. There is presently no standard or an entirely conventional platform that addresses the intricacies of emphasizing networks and communications technologies while also providing a simple branding service for a variety of applications. The generated services may be incompatible with one-of-a-kind communication and implementation environments. Furthermore, powerful web service methods and solutions for object naming must be developed to disseminate IOT science.

4.9 Conclusion and Future Work

Industry 4.0 represents a significant change in how we live and interact. A new era of human development has begun due to massive technological developments similar to the Industrial Revolution. These advances are integrating the physical, and digital worlds in a way that also carries significant risks. Revolutions are driving us to rethink how nations evolve. Industry 4.0 offers opportunities for all citizens from all socioeconomic levels and countries. There is real hope in going beyond technology and creating ways to enable as many people as possible to have a good impact on their families, organizations, and communities. Achieving resource-efficient and user-friendly production solutions require intelligent and precise equipment. Industry 4.0 is expected to benefit from new technologies and applications by increasing production and providing customized products on demand. We begin by outlining the many new concepts and definitions of Industry 4.0 as observed from the perspectives of numerous industry practitioners and experts.

The term "Industry 4.0" first appeared a few years ago in response to the concept of Industrial Revolution 4.0. It has become a subject of increasing concern to policymakers and organizations due to the impact of the pandemic, there is focus on topics such as sustainability and resilience, and a renewed focus on human needs. Major industries around the world are looking for headlines about cutting-edge technologies that can help them streamline operations and reduce expenses. Executives at logistics companies do the same to better prepare their companies for the future. After all, the logistics business has traditionally relied on manual operations and data storage, and new technological solutions help it a lot. Cutting-edge technology refers to technical equipment and processes. We need to continue researching on cutting-edge technologies.

References

1. Paschek, D., Mocan, A., & Draghici, A. (2019, May). *Industry 5.0—The expected impact of the next industrial revolution.* In *Thriving on Future Education, Industry, Business, and Society, Proceedings of the Make Learn and TIIM International Conference*, Piran, Slovenia (pp. 15–17). http://www.toknowpress.net/ISBN/978-961-6914-25-3/papers/ML19-017.pdf
2. Almada-Labo, F. (2017). Six benefits of Industrie 4.0 for businesses. Retrieved from: https://www.controleng.com/articles/six-benefits-of-industrie-4-0-for-businesses/
3. Atwell, C. (2017). Yes, Industry 5.0 is already on the Horizon. Retrieved from: https://www.machinedesign.com/industrial-automation/yes-industry-50-already-horizon
4. Gartner (2017). IT is glossary; digitalization. Retrieved from: https://www.gartner.com/itglossary/digitalization. Retrieved: 24.04.2018
5. EESC (2018). Industry 5.0. Retrieved from: https://www.eesc.europa.eu/en/agenda/ourevents/events/industry-50
6. Kospanos, V. (2017). Industry 5.0—Far from science fiction (pt. 2). Retrieved from: http://www.pnmsoft.com/industry-5-0-far-science-fiction-pt-2/
7. Köhler-Schute, C. (2016). *Digitalisierung und Transformation in Unternehmen: Strategien und Konzepte, Methoden und Technologien, Praxisbeispiele*, Berlin: Germany KS-Energy-Verlag.
8. Lewis, A., (2016) Guide to Industry 4.0 & 5.0. Retrieved from: https://blog.gesrepair.com/category/manufacturing-blog/
9. Marr, B. (2018). The 4th industrial revolution is here—Are you ready?. Retrieved from: https://www.forbes.com/sites/bernardmarr/2018/08/13/the-4th-industrial-revolution-is-here-areyou-ready/#2b6b3e2628b2
10. Mohelska, H., & Sokolova, M. (2018). Management approaches for industry 4.0—The organizational culture perspective, *Technological and Economic Development of Economy*, VGTU Press.

11. Newman, D. (2017) Four digital transformation trends driving Industry 4.0. Retrieved from https://www.forbes.com/sites/danielnewman/2018/06/12/four-digital-transformation-trends-drivingindustry-4-0/#1bf42316604a. Retrieved 03.01.2019
12. Paschek, D., et al. (2017). *Business process as a service—A flexible approach for it service management and business process outsourcing, Management Challenges in a Network Economy, Proceedings of the Make Learn Conference 2017.*
13. Pearce, R. (2017). How to be part of the fifth industrial revolution. Retrieved from: https://www.inmarsat.com/blog/how-to-be-part-of-the-fifth-industrial-revolution/
14. Rada, M. (2018). Industry 5.0 definition. https://medium.com/@michael.rada/industry-5-0-definition-6a2f9922dc48
15. Rundle, E. (2017). The 5th industrial revolution, when it will happen and how. Retrieved from: https://devops.com/5th-industrial-revolution-will-happen/
16. Sachsenmeier, P. (2016). Industry 5.0—The relevance and implications of bionics and synthetic biology. *Engineering* 2. Retrieved from: https://www.journals.elsevier.com/engineering
17. Scanlon, S. (2018). Now prepare for the 5th industrial revolution. Retrieved from: https://gadget.co.za/now-prepare-for-the-5th-industrial-revolution/
18. Shelzer, R. (2017). What is Industry 5.0—And how will it affect manufacturers?. Retrieved from: https://blog.gesrepair.com/industry-5-0-will-affect-manufacturers/
19. Urbach, N. (2018). *Digitalization Cases, How Organizations Rethink Their Business for the Digital Age*, Cham, Switzerland: Springer Nature Switzerland AG.
20. Ustundag, A., & Cevican, E. (2018). *Industry 4.0: Managing The Digital Transformation*, Cham, Switzerland: Springer Nature Switzerland AG.
21. Wang, K., Wang, Y. Strandhagen, J. O., & Yu, T. (2016). *Advanced Manufacturing and Automation V*, WIT Press.
22. Seungjin, L., Abdullah, A., & Jhanjhi, N. Z. (2020). A review on honeypot-based botnet detection models for smart factory. *International Journal of Advanced Computer Science and Applications, 11*(6), 418–435. https://ieeexplore.ieee.org/abstract/document/9214512/
23. Lee, S., Abdullah, A., Jhanjhi, N., & Kok, S. (2021). Classification of botnet attacks in IoT smart factory using honeypot combined with machine learning. *PeerJ Computer Science, 7*, e350. https://seap.taylors.edu.my/file/rems/publication/109566_8816_1.pdf
24. Gaur, L., Afaq, A., Solanki, A., Singh, G., Sharma, S., Jhanjhi, N. Z., ... & Le, D. N. (2021). Capitalizing on big data and revolutionary 5G technology: Extracting and visualizing ratings and reviews of global chain hotels. *Computers & Electrical Engineering, 95*, 107374. https://www.sciencedirect.com/science/article/abs/pii/S0045790621003438
25. Jayakumar, P., Brohi, S. N., & Jhanjhi, N. Z. (2021). Artificial intelligence and military applications: Innovations, cybersecurity challenges & open research areas. https://www.preprints.org/manuscript/202108.0047/v1
26. Ullah, A., Ishaq, N., Azeem, M., Ashraf, H., Jhanjhi, N. Z., Humayun, M., ... & Almusaylim, Z. A. (2021). A survey on continuous object tracking and boundary detection schemes in IoT assisted wireless sensor networks. *IEEE Access, 9*, 126324–126336. https://ieeexplore.ieee.org/abstract/document/9529170

27. Sujatha, R., Chatterjee, J. M., Jhanjhi, N. Z., & Brohi, S. N. (2021). Performance of deep learning vs machine learning in plant leaf disease detection. *Microprocessors and Microsystems, 80*, 103615. https://www.sciencedirect.com/science/article/abs/pii/S0141933120307626
28. Sujatha, R., Aarthy, S. L., Chatterjee, J., Alaboudi, A., & Jhanjhi, N. Z. (2021). A machine learning way to classify autism spectrum disorder. *International Journal of Emerging Technologies in Learning (IJET), 16*(6), 182–200. https://www.learntechlib.org/p/219973/
29. Tri, N. M., Hoang, P. D., & Dung, N. T. (2021). Impact of the industrial revolution 4.0 on higher education in Vietnam: Challenges and opportunities. *Linguistics and Culture Review, 5*(S3), 1–15. http://lingcure.org/index.php/journal/article/view/1350
30. Hussain, A. (2019). Industrial revolution 4.0: Implication to libraries and librarians. *Library hi tech news*. https://www.emerald.com/insight/content/doi/10.1108/LHTN-05-2019-0033/full/html
31. Lase, D. (2019). Education and industrial revolution 4.0. *Jurnal Handayani Pgsd Fip Unimed, 10*(1), 48–62. https://jurnal.unimed.ac.id/2012/index.php/handayani/article/view/14138
32. Darmaji, D., Kurniawan, D., Astalini, A., Lumbantoruan, A., & Samosir, S. (2019). Mobile learning in higher education for the industrial revolution 4.0: Perception and response of physics practicum. https://www.learntechlib.org/p/216574/
33. Dewi, M. V. K., & Darma, G. S. (2019). The role of marketing & competitive intelligence in industrial revolution 4.0. *Jurnal Manajemen Bisnis, 16*(1), 1–12. https://publisher.uthm.edu.my/ojs/index.php/JTET/article/view/3208
34. Robandi, B., Kurniati, E., & Sari, R. P. (2019, April). *Pedagogy in the era of industrial revolution 4.0*. In *8th UPI-UPSI International Conference 2018 (UPI-UPSI 2018)* (pp. 38–46). Atlantis Press. https://www.atlantis-press.com/proceedings/upiupsi-18/55916302
35. Rymarczyk, J. (2020). Technologies, opportunities and challenges of the industrial revolution 4.0: Theoretical considerations. *Entrepreneurial Business and Economics Review, 8*(1), 185–198. https://www.ceeol.com/search/article-detail?id=976200
36. Ellitan, L. (2020). Competing in the era of industrial revolution 4.0 and society 5.0. *Jurnal Maksipreneur: Manajemen, Koperasi, dan Entrepreneurship, 10*(1), 1–12. https://www.sciencedirect.com/science/article/pii/S2090447919301157
37. Sherwani, F., Asad, M. M., & Ibrahim, B. S. K. K. (2020, March). *Collaborative robots and industrial revolution 4.0 (ir 4.0)*. In *2020 International Conference on Emerging Trends in Smart Technologies (ICETST)* (pp. 1–5). IEEE. https://ieeexplore.ieee.org/abstract/document/9080724
38. Farida, I., Setiawan, R., Maryatmi, A. S., & Juwita, M. N. (2020). The implementation of E-government in the industrial revolution era 4.0 in Indonesia. *International Journal of Progressive Sciences and Technologies, 22*(2), 340–346. http://ijpsat.es/index.php/ijpsat/article/view/2165
39. Harahap, N. J., & Rafika, M. (2020). Industrial revolution 4.0: And the impact on human resources. *Ecobisma (jurnal ekonomi, bisnis dan manajemen), 7*(1), 89–96. https://jurnal.ulb.ac.id/index.php/ecobisma/article/view/1545

40. Wong, B. K. M., & Hazley, S. A. S. A. (2020). The future of health tourism in the industrial revolution 4.0 era. *Journal of Tourism Futures*. https://www.emerald.com/insight/content/doi/10.1108/JTF-01-2020-0006/full/html
41. Humayun, M., Jhanjhi, N. Z., & Almotilag, A. (2022). Real-time security health and privacy monitoring for Saudi highways using cutting-edge technologies. *Applied Sciences*, *12*(4), 2177. https://www.tandfonline.com/doi/full/10.1080/08839514.2022.2037255
42. Humayun, M., Almufareh, M. F., & Jhanjhi, N. Z. (2022). Autonomous traffic system for emergency vehicles. *Electronics*, *11*(4), 510. https://www.mdpi.com/2079-9292/11/4/510
43. Li, J., Goh, W. W., Jhanjhi, N. Z., Isa, F. B. M., & Balakrishnan, S. (2021). An empirical study on challenges faced by the elderly in care centres. *EAI Endorsed Transactions on Pervasive Health and Technology*, e2. http://eprints.eudl.eu/id/eprint/5165/
44. Khan, A., Jhanjhi, N. Z., Humayun, M., & Ahmad, M. (2020). The role of IoT in digital governance. In *Employing Recent Technologies for Improved Digital Governance* (pp. 128–150). IGI Global. https://www.igi-global.com/chapter/the-role-of-iot-in-digital-governance/245979
45. Humayun, M., Jhanjhi, N. Z., Hamid, B., & Ahmed, G. (2020). Emerging smart logistics and transportation using IoT and blockchain. *IEEE Internet of Things Magazine*, *3*(2), 58–62. https://ieeexplore.ieee.org/abstract/document/9125435
46. Singh, A. P., Pradhan, N. R., Luhach, A. K., Agnihotri, S., Jhanjhi, N. Z., Verma, S., ... & Roy, D. S. (2020). A novel patient-centric architectural framework for blockchain-enabled healthcare applications. *IEEE Transactions on Industrial Informatics*, *17*(8), 5779–5789. https://ieeexplore.ieee.org/abstract/document/9259231
47. Kumar, M. S., Vimal, S., Jhanjhi, N. Z., Dhanabalan, S. S., & Alhumyani, H. A. (2021). Blockchain based peer to peer communication in autonomous drone operation. *Energy Reports*, *7*, 7925–7939. https://expert.taylors.edu.my/file/rems/publication/109566_6018_1.pdf
48. Kumar, M. S., Vimal, S., Jhanjhi, N. Z., Dhanabalan, S. S., & Alhumyani, H. A. (2021). Blockchain based peer to peer communication in autonomous drone operation. *Energy Reports*, *7*, 7925–7939. https://www.sciencedirect.com/science/article/pii/S2352484721006752
49. Rajmohan, R., Kumar, T. A., Pavithra, M., Sandhya, S. G., Julie, E. G., Nayahi, J. J. V., & Jhanjhi, N. Z. (2020). Blockchain: Next-generation technology for Industry 4.0. *Blockchain Technology*, 177–198. https://www.researchgate.net/profile/T-Ananth-Kumar/publication/346792929_Blockchain_Technology/links/601a0281a6fdcc37a8fc1ef5/Blockchain-Technology.pdf

5

Secure IoT Protocol for Implementing Classified Electroencephalogram (EEG) Signals in the Field of Smart Healthcare

S. Saravanan

SASTRA (Deemed to be University), Kumbakonam, India

M. Lavanya

SASTRA (Deemed to be University), Thanjavur, India

Chandra Mouli Venkata Srinivas

BVC Engineering College, Odalarevu, India

M. Arunadevi

Cambridge Institute of Technology, Bangalore, India

N. Arulkumar

CHRIST (Deemed to be University), Bangalore, India

CONTENTS

DOI: 10.1201/9781003203087-5

5.1 Introduction

Today's technology focuses on the Industry 4.0 revolution with great expectations in the domain of smart automation, trusted information storage, huge data processing, massive production processing, and secure healthcare. Particularly in the healthcare sector, the Industry 4.0 revolution focuses on large-scale data processing, digitization, diagnosing, monitoring, secure data transfer, and trusted communication between patients and caretakers. This Industry 4.0 integrates various technologies like the Internet of Things (IoT), Artificial Intelligence (AI), Machine to Machine (M2M) communication, cloud computing and storage, data analysis, and secure data transfer [1]. This framework helps to develop a better healthcare system, understanding the patient's problem and recovery methods.

This chapter focuses on Electroencephalography (EEG) signals, which are involved in various research fields like biomedical, bioengineering, neural science, and brain science. It measures the potential of the membrane and neuron strength through various electrodes. These signals help to analyze seizure detection, muscle weakness, body coma condition, brain death, non-invasiveness, stroke, and sleep disorders. EEG signals are received from the skull through various electrodes to measure brain activity. The electrode positions in the skull are followed by the International 10–20 standard system. All the EEG signals are in the analog domain and need to be sampled by ADC (Analog to Digital Converter). The received EEG signals are classified into delta, theta, alpha, beta, and gamma frequencies. Depending upon the range of low frequency to high frequency, it indicates the patient condition from relax mode (asleep) to severe mode (brain dead). The main advantage of EEG signals is supporting high temporal resolution along with low-cost equipment.

An EEG signal is one of the major bioinformation which plays a vital role in the Tele-healthcare system. The need for increased patients and fewer clinical facilities forces Tele-healthcare into an unavoidable technology. Tele-healthcare promises low cost with improved healthcare quality. This technology is more comfortable and adaptable to IoT for providing a suitable solution to health information.

The basic structure of IoT is all about extending the internet beyond the embedded systems and smart mobiles device. It has been predicted that by 2025 over 70 billion things (devices) will be connected to the internet connection by generating 79.4 Zettabytes (ZB) of information. IoT-based technology has transformed the realm of Tele-healthcare by providing compatible support to health monitoring services. In the current scenario, IoT-based Tele-healthcare technology has a lot of preferences and is considered one of the emerging technologies. This technology helps to receive the human body signals through various biosensors to observe, monitor, and make diagnoses.

It integrates various devices such as sensors, actuators, and communication protocols along with internet connections to support healthcare-related issues. With this technology, the patients do not need to visit a hospital for medical check-ups to know about their health problems. They can get all the medical support and treatment from physicians through a remote IoT environment. This approach helps for an efficient process and better comfort and provides a suitable solution for health information. This system also has various advantages like privacy assurance, cost efficiency, easy accessibility, and improved healthcare quality.

Various IoT architectures were proposed by researchers and developers to full fill the need of the application. Basic IoT architecture is considered in three layers [2,3]. They are the Perception layer (base), the Network layer (middle), and the Application layer (top). In the base layer, all the real-time biosignals are collected with help of various sensors. In the middle layer, it connects the sensor data with various gateway devices and communication protocols. This middle layer also improves the efficiency of biosignals with low power consumption. In the top layer, a personalized biosignal integrates with user applications and secure parameters.

Intermediate IoT architecture contains four layers of the structure by including a support layer along with a three-layer IoT architecture. The fourth layer helps with security and authentication purposes. It is mainly used for confirming secure data along with trusted users and protected connections. This layer also checks Denial of Service (DoS) attacks and malicious attacks [4]. Advanced IoT architecture consists of five layers with the inclusion of the processing layer and the business layer. In the processing layer, it supports the elimination of unnecessary data or extra data. This layer also influences exhaustion and malware attacks. In the business layer, it acts as a manager for the overall application and is also responsible to control and manage the particular application. It is also vulnerable to business logic attacks [5] and zero-day attacks.

IoT-based environment receives the real-time EEG signals and processes through a machine-learning-based classification algorithm. It helps for a better solution to filter unwanted bioinformation and gives the needed information. Various classification algorithms are available for the data filtering process. This chapter focuses on the Support Vector Machine (SVM) classification algorithm. This algorithm is basically known for learning the statistical theory and giving a better solution for two-class linear approaches. EEG signal is closely associated with the monitoring symptoms of headache, weakness, paralysis, poor eye movement, and seizure.

Secure healthcare-based diagnosis [6] is more vulnerable to security threats and middle-man attacks. It is more important to address this issue by providing better security in terms of bioinformation store, communication protocol, and integrity. One of the solutions for secure communication can be achieved by secure IoT protocol. This protocol is used for information transfer along with secure communication [7]. Message Queue Telemetry Transport (MQTT) protocol is one of the major IoT protocols used in the IoT application layer. This protocol works on a lightweight concept and gives more compatibility to integrate with the IoT environment. This chapter highlights the importance of secure MQTT protocol by providing username/password, implementing secure Transport Layer Security (TLS)/Secure Sockets Layer (SSL), and Arduino IoT environment.

This chapter provides an overall observation and process flow of secure healthcare in secure IoT protocol. This chapter helps the audience to know about the progress of EEG signals in the IoT environment, the classification of received EEG signals using machine learning algorithm, and highlights the importance of secure MQTT IoT protocol by providing secure communication. Finally, cloud-based arrangements help for secure MQTT protocol to achieve better data storage and processing of bioinformation.

5.2 Existing Methods

Various research methods have been targeted toward Industry 4.0 and IoT-based secure healthcare systems. This section gives an elaborated view of earlier existing methods.

In IoT architecture, lightweight-based cryptography [8] was discussed in terms of distributed security and resource minimization. IoT-based embedded security [9] was focused on as a framework of security in data communication and data authentication. In [10], the security of IoT was considered in encrypted and hash-based data processing. This method works under the perception layer and activated security in the base layer. Software-Defined Network (SDN)-based IoT method was discussed in [11] by providing better performance with less cost and minimized hardware design. This method gave security for both IoT controller and agent. Two layers prevention system was elaborated in [12] for wireless sensor networks. This method helps to give better security in the network layer of IoT architecture. In [13], a heterogeneous-based IoT approach was proposed along with an integrated platform. This article gave an in-depth view of heterogeneous design in IoT applications.

Industry 4.0 in AI and IoT challenges and its security issues were focused on in [14] in terms of preventing malicious attacks, unauthorized accessing, and future challenges. In [15], an elaborated survey on trust management was discussed for IoT-based applications. Privacy was given more importance in distributed-based IoT in [16]. This paper also discussed security involvement and its countermeasure in distributed IoT. Various research directions for IoT were explained in [17] and an IoT security survey was highlighted in [18].

More research work is also carried over in the field of Tele-healthcare in association with IoT-based environments. In [19], a broad survey was done in terms of medical records by using data mining technology. The paper elaborates on the use of medical records in electronic format. Smart and secure health monitoring was discussed in [20] along with various IoT sensors. The collected healthcare information is processed through cloud computing. A review related to cost reduction of surgical was explained in [21]. This paper assists to understand the surgical information and its various secure parameters. In [22], IoT-based medical parameters were described along with various privacy and security issues. This paper provides an elaborated review of the medical IoT environment.

Secure healthcare information gives more compatibility to process with IoT environment. This chapter mainly focuses on EEG-based secure healthcare information and identified a few existing research works. In [23], EEG signals are observed with least cost headset and utilized IoT environment. A machine learning algorithm was used to classify the EEG signals in terms of alpha and beta signals in [24]. Patient's mental parameter was observed with help of EEG signals and the signals were classified with SVM algorithm [25]. Industry 4.0 healthcare and its various possible methods are observed in [25–29]. These papers discussed the importance of Industry 4.0 in India by TISM, visions for smart healthcare, IoT, Big Data, Cloud computing, and various applications in the medical field.

The importance of secure IoT protocol was observed from various existing methods. In [30], secure MQTT was described with Fuzzy logic to observe DoS attacks. A generic IoT model for MQTT protocol was focused on in [31] with a secure framework. In [32], security-based test parameter was discussed along with MQTT architecture. Still, IoT-based Tele-healthcare does not have matured architecture and design for transforming secure bioinformation over the internet connection.

5.3 EEG Signals in IoT Environment

5.3.1 Basics of EEG Signals

Electroencephalogram (EEG) refers to the human brain signal in the form of electrophysiological, which helps to record the electrical signals through sensors. This signal is a non-invasive method to record voltage fluctuations of the brain neurons and assist to know the real-time behavior of the human brain from the scalp. Basically, the EEG signal is measured in microvolt (μV) and contains electrical characteristics such as amplitude, waveform shape, and frequency.

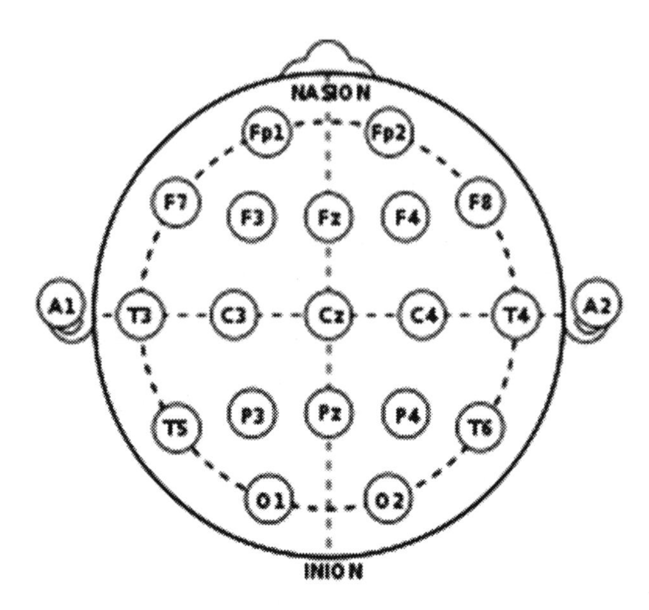

FIGURE 5.1
Location of electrodes to receive EEG signal [33].

EEG electrodes follow the International benchmark system, which consists of 10/20 electrode position placed on the specific location of the scalp between Nasion-Inion and fixed points. The representation of 10/20 indicates the actual distance between the front and back or left and right sides of the skull with 10% or 20% of the distance. Basically, electrodes are small metal discs and are usually covered with a silver chloride coating. Each electrode is represented with a letter such as Front Pole (Fp), Central (C), Temporal (T), Parietal (P), Occipital (O), and Zero (Z) in the midline as shown in Figure 5.1. A few electrodes are also represented in odd numbers for the left hemisphere and even numbers for the right hemisphere.

EEG waveform frequencies are classified into delta (0–3Hz), theta (4–7Hz), alpha (8–15Hz), beta (16–31 Hz), and gamma (>32 Hz) as shown in Figure 5.2. EEG signals help to diagnose brain activities such as coma, death, muscle weakness, epilepsy, stroke, tumor, and sleep disorders.

The Delta (0–3Hz) consists of the highest amplitude and slowest wave. It is also known as slow-wave sleep in adults and babies. It helps to diagnose subcortical injuries and deep midline lesions. The Theta (4–7Hz) signals are associated with young children and also have a state of meditation, relaxation, and creativity. It supports the diagnosis of diffuse disorder or metabolic encephalopathy. The Alpha (8–15Hz) is also called Posterior Dominant Rhythm (PDR) and is seen on both sides of the head. It identifies the relaxation of the eyes and mental exertion. The Beta (16–31 Hz) is seen usually on both sides in a symmetrical distribution. It is associated with

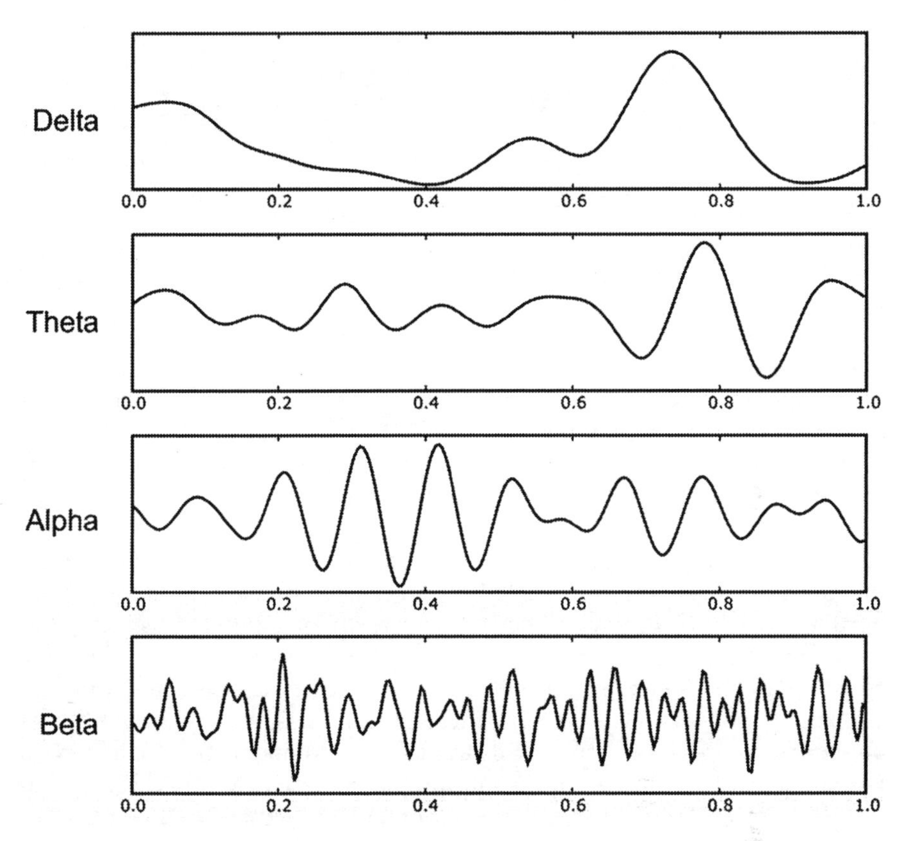

FIGURE 5.2
Classification of EEG waveform frequencies [24].

more concentration, busy thinking, and anxious thinking. The Gamma (>32 Hz) signal is associated with the binding of various populations of neurons together into a common network for cognitive function. The EEG frequencies are also broadly classified into different properties like Rhythmic (constant frequency), Arrhythmic (not stable frequency), and Dysrhythmic (Rarely seen patient frequency).

5.3.2 IoT Environment

In the IoT environment, Arduino [34] or Raspberry Pi [35]-based design boards are very popular. Arduino contains an Atmel family microcontroller with an inbuilt boot loader for plug-and-play embedded programming. Its open-source software comes with an Integrated Design Environment (IDE), which helps to write a program and debug and dump the program into

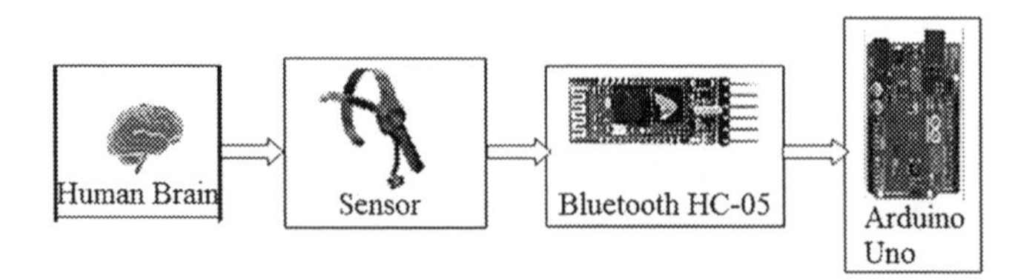

FIGURE 5.3
Block diagram of IoT environment.

Arduino hardware. This arrangement works with a Serial Communication window through which we can easily get the serial data from the board.

In Raspberry Pi, it is considered a full-fledged computer by having its own processor, graphic display unit, HDMI, USB, and memory. It works under Raspberry Pi OS (formerly called Raspbian) under Linux. It also supports Python and Scratch as the main programming languages, with support for many other languages.

The overall block diagram of the IoT environment is shown in Figure 5.3. Initially, EEG signals were collected from the patient in real time to observe the brain activity. MindWave Headset, which works in Bluetooth connection helps in receiving EEG signals. This product is from NeuroSky, and it is very easy to place in the human skull. Received signals were transmitted through a Bluetooth connection. To receive these EEG signals, it is important to use Bluetooth devices compatible with the IoT environment. Bluetooth device HC-05 supports an IoT environment and is connected to an Arduino Uno board. Through Arduino programming, all the EEG signals were captured in terms of Delta, Theta, Alpha, Beta, and Gamma frequency.

5.4 Classification Method

A received EEG signal is analyzed with Machine Learning (ML) algorithm and very particularly focuses on the SVM classification algorithm. This SVM algorithm is basically known as statistical-based learning theory, which gives a better solution in hyperplane for two-class linear approaches. This algorithm also helps to solve linear and non-linear-based issues in the dataset. The proposed stages are given in Figure 5.4.

In the preprocessing stage, raw EEG signals are filtered to get clear data. A high pass filter is used to remove DC components and drifts. A low pass filter is also used to remove high-frequency signals which occur in the gamma range. A single EEG signal is capable of extracting various brain signals like eye open/close, strong emotional thoughts, and active minds. In the Feature Extraction stage, the hidden information in EEG signals are extracted

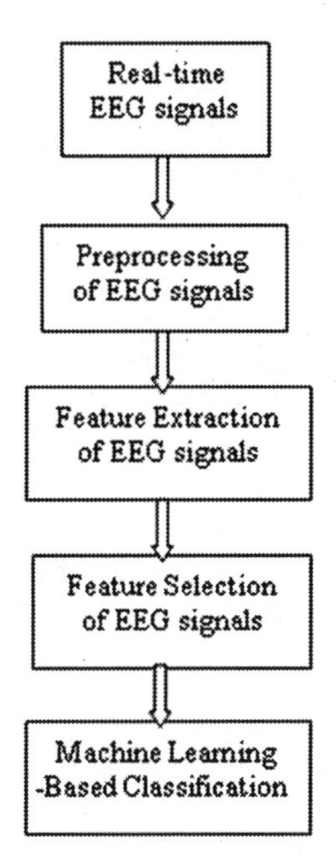

FIGURE 5.4
Real-time EEG signal classification stages.

with a complex automatic processing algorithm. There are various methods supporting EEG signal extractions in terms of time domain and frequency domain. In the time domain, entropy, mean, and standard deviation are considered. In the frequency domain, Fourier transform wavelets are used.

In Feature selection, specific EEG signal information gets more important. For example, relax and stress states are common in the human brain. Depending upon the above brain state, feature selection helps to find out among more differentiation between these two conditions. In the Classification stage, it recognizes our required features from one class/condition (relax) or to another class/condition (stress). In this chapter, SVM-based machine learning classification is considered.

This machine learning algorithm supports finite-dimensional vector space to represent a feature of a particular object. The main objective of this SVM algorithm is to train the model that was assigned new hidden objects into a particular required category. It is considered non-probabilistic because the new object is fully determined by its location and there is no stochastic element involved. This SVM algorithm contains various advantages like high dimensionality, memory efficiency, and versatility.

This algorithm has various advantages like effective utilization of memory, improved high dimensional space, a clear margin in the separation of different classes, more suitability for multiple feature datasets, and support of using common kernel functions.

The following steps are considered in SVM-based classification algorithm in Python-based processing method. Sample EEG data set is available in [36].

Step 1: import pandas as pd, include dataset
Step 2: dataset shape (dimension size)
Step 3: drop the unnecessary dataset field
Step 4: split the dataset into train and test before the SVM algorithm
Step 5: import SVM along with the train dataset
Step 6: Use predict values and evaluate the SVM model
Step 7: Final result after SVM classification

5.5 Secure IoT Protocol

IoT-based applications are booming up along with various security aspects. It is mandatory to observe five basic parameters, Data Identity, Authorization, Confidentiality, Data Integrity, and Accountability, in the development of secure IoT-based applications. In general, MQTT works with the concept of a Client and a Broker network. Any processing device like a microprocessor, microcontroller, and server can act as a client module. A broker network receives the real-time information from the client and it is also responsible for identifying continuous connections from the client module.

Data identity is considered an initial stage, where the client is initiating a request to the broker. Every client contains a proper Universal Unique Identifier (UUID), which helps the broker to identify the data from the corresponding client. Data can also identify the client through the network Media Access Control (MAC) address. Thus, the UUID and MAC address helps in identifying the authorized access in data transfer. Authorization plays a vital role in secure data accessing on communication channels. It also provides authentication for IoT data to monitor and access real-time information.

Confidentiality allows only authorized users, devices, and entities. It controls the transmission of message transfer and access. It also helps to protect the private information, accessing key, and credentials from unauthorized access. Data integrity provides support to various data requirements for a particular IoT application. It protects all the original data along with its supporting data from data tampering. As an example, Telemedicine data have more protection than a Smart irrigation system. Accountability helps to improve the toughness of IoT information in terms of the system environment, security, and communication protocol.

MQTT protocol [37] is considered in this chapter. It is a lightweight protocol and very easy to implement in software for better data transmission. It basically works as a server and a client concept with low network usage and low power consumption. The server (broker) is responsible for handling the client's requests for sending or receiving data between each other. This arrangement is also known as Publish and Subscribe. In the Publish concept, a client wants to send data to the broker and in Subscribe, a client wants to receive data from the broker.

The basic arrangement of the MQTT protocol packet format is given in Figure 5.5. The field length is considered as 8 bit for processing the information. In byte 1, message type, DUP (duplicate flag), and Quality of Service (QoS) level bit were considered. In byte 2, the remaining length (1–4 bytes) was used to complete the communication. Byte 3 to byte n are used for variable-length header and byte n+1 to byte m contains message payload. When a connection is activated between client and broker, the subscriber needs to receive one or many topics from the client.

Secure MQTT provides the authentication to confirm the truth of an attribute of a single data or information. It provides username and password fields in the communication message or in the connect message. Thus, it activates the client to have the possibility to send an authenticated username and password when connected to an MQTT broker. Then, the broker will evaluate the authentication credentials on the implemented mechanism and return various codes like 0 – accepted, 1 – bad username and password, and 2– not authorized. The client identifier is also authenticating MQTT communication by providing a unique client identifier, serial number, and unique MAC identification.

Figure 5.6 shows the basic arrangements of a communication protocol for an IoT-based environment. The base layer is considered a network/ link layer, which supports Bluetooth Low Energy (BLE) /IEEE 803.15.4.

Field Length (bits)	0	1	2	3	4	5	6	7
Byte 1	Message Type				DUP	QoS Level		Retain
Byte 2	Remaining length (1 to 4 bytes)							
Byte 3	Variable length header (Optional)							
—								
Byte n								
Byte n+1	Variable length message payload (Optional)							
—								
Byte m								

FIGURE 5.5
Basic MQTT packet format.

Layer Stages	Protocol Usage
Application Layer	MQTT / CoAP
Transport Layer	TCP / UDP
Internet Layer	IPv6 / 6LoWPAN
Network / Link layer	BLE / IEEE 803.15.4

FIGURE 5.6
Communication protocol for IoT-based environment.

The second layer is known as the internet layer, which helps to create IPv6 protocol over Low Power Wireless Personal Area Network (LoWPAN). The third layer supports TCP and UDP protocols in the transport layer. The top layer provides MQTT/CoAP protocol service to the application layer. TCP supports for MQTT protocol and UDP helps the CoAP protocol to run over in the IoT transport layer.

Various advantages of the MQTT protocol are listed below.

- More effective in data/information distribution
- Helps to increase the scalability
- Works in lightweight processing
- Provides better security based on a permission-based approach
- Reduces the duration of the development phase
- Publisher- and Subscriber-based protocol supports large data communication along with less bandwidth.

5.5.1 Basic MQTT Protocol

The following procedure shows the basic structure of the MQTT protocol in Python language as shown in Figure 5.7.

```
pip --version

pip 19.3.1 from /usr/local/lib/python3.6/dist-packages/pip (python 3.6)

pip install paho-mqtt

Collecting paho-mqtt
  Downloading https://files.pythonhosted.org/packages/32/d3/6dcb8fd14746fcde6a556f932b5de8bea8fedcb85b3a092e0e986372c0e7/paho-mqtt-1.5.1.tar.gz (101kB)
     |████████████████████████████████| 102kB 4.2MB/s
Building wheels for collected packages: paho-mqtt
  Building wheel for paho-mqtt (setup.py) ... done
  Created wheel for paho-mqtt: filename=paho_mqtt-1.5.1-cp36-none-any.whl size=61544 sha256=76c28b4473d6cd2e79eb42194794f653d0fe4ed02fa1b96e50ce2cdf93f3d9b7
  Stored in directory: /root/.cache/pip/wheels/75/e2/f8/75942b19b4d135605e58dfe88fba82253b14d636aabf76904b
Successfully built paho-mqtt
Installing collected packages: paho-mqtt
Successfully installed paho-mqtt-1.5.1
```

FIGURE 5.7
MQTT protocol in Python language.

The pip is basically a tool, which allows the user to manage and include additional Python libraries. It acts as a packet manager for Python language. pahomqtt client is one of contains e free public MQTT broker with various methods like connect(), disconnect(), subscribe(), unscribe(), and publish(). In MQTT Python-based publisher code contains the inclusion of paho.mqtt.client, broker address, client.connect, client.loop, client.publish (text), client.disconnect, and client.loopstop. A few examples of free online access MQTT brokers are www.broker.hivemq.com [38], www.test.mosquitto.org [39], and www.iot.eclipse.org [40]. In MQTT, Python-based subscriber code contains import of paho.mqtt.clinet, broker address (same as provided in publisher code), def_on_connect, def_on_message, client.connect, client_on_connect, client_on_message, and client.loopforever.

5.5.2 MQTT Protocol with Username and Password

The same code along with username and password gives single-layer security. The following information provides single-layer security for the MQTT protocol in terms of username and password. This authentication connects with the transport layer and the application layer of the MQTT protocol. In the transport layer, client-based certification is used, and in the application layer, user and password protect the MQTT protocol.

Authenticated MQTT protocol consists of the following Python syntax:

```
User = "username"
Password="password"
Client.username_pw_set ("username", "password")
```

The above syntax is considered as a single protection with a simple entity. It helps to identify the authenticated person, device, and application by verifying the username and password. When this single layer security is inbuilt with MQTT protocol, the corresponding MQTT broker evaluates the authentication and response like connection accepted or connection failed due to invalid username and password. Thus, it is very easy to provide an inbuilt username and password in MQTT protocol for secure data transmission.

5.5.3 MQTT Protocol in IoT Arduino Environment

One of the common IoT environments is known as Arduino hardware [34] which also supports MQTT protocol along with username and password protections. Arduino supports Ethernet network and allows the user to connect to the internet. It is also possible to use Ethernet and MQTT library to develop MQTT client-server. MQTT libraries are not by default in the

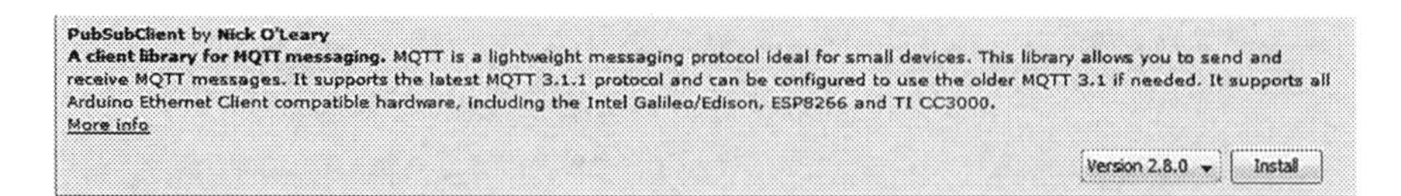

FIGURE 5.8
MQTT library installation.

Arduino environment and it is necessary to install them through navigation to the Arduino IDE sketch as shown in Figure 5.8. The steps are:

Step 1: Arduinouno sketch > Include the new libraries > Manage libraries (for searching MQTT libraries – PubSubClient)

Step 2: Create a new file > include libraries for SPI.h, Ethernet.h, and PubSubClient.h.
SPI library (Serial Peripheral Interface) is used for hardware communication over shorter distances and for quicker responses. It is also used for the Ethernet support library. PubSubClient library is used for activating the MQTT protocol.

Step 3: Define MQTT specification in terms of IP address, MAC address, and server configuration
The IP address is considered for an Ethernet connection and the MAC address contains a unique value. Server configuration is a simple design and possible to use free of MQTT broker.

Step 4: In Void setup, include serial begin for debug process, Ethernet begin, Set MQTT server, Attempt to connect the server, and Establish subscribe.

Step 5: In the Void loop, include the top loop, and attempt to publish and subscribe to receiver with a message.

5.5.4 MQTT Protocol with Mosquitto Broker

Free online access of MQTT brokers also helps in secure IoT connections. The following steps are considered to achieve a secure MQTT protocol with the Mosquitto broker [39]. It is an open-source message broker service that uses the MQTT protocol. It also helps to send and receive messages in an IoT environment.

Step 1: Download mosquitto from https://mosquitto.org/

Step 2: Download openssl from http://slproweb.com/

Openssl tool [41] is used to generate private key generation and Certificate Signing Request (CSR). It is also considered an overall cryptography library with open-source applications.

Step 3: Copy openssl32\bin folder to the mosquitto folder
Step 4: Copy required libeay32.dll files to the mosquitto folder
Step 5: Copy required ssleay32.dll files to the mosquitto folder
Step 6: Download pthreadVC2.dll from ftp://sources.redhat.com/

PthreadVC2.dll [42] is considered a Dynamic Link Library (DLL) file developed by Open Source Software, which is referred to as an essential system file of the Windows OS. It usually contains a set of procedures and driver functions, which may be applied by Windows.

Step 7: Copy pthreadVC2.dll to the mosquitto folder
Step 8: Test subscriber/server using mosquitto_sub -v -t 'text'
Step 9: Test publisher/client mosquitto_pub -t 'text' -m 'Condition'

5.5.5 MQTT Protocol with SSL/TLS Configuration

The TLS/SSL configuration method provides security in MQTT Protocol [43]. The following information steps are very common in general and for more accurate certification authority, the user needs to refer to the corresponding broker information. These steps are very particular in SSL/TLS configuration for mosquitto broker arrangements.

Step 1: Create a new Certificate Authority (CA) key pair
This step helps to declare the encryption process along with the private key size

Step 2: Create CA certificate along with key pair which is created in step 1
This step creates X.509 certificate along with a private key

Step 3: Create a broker key pair which is not protected by a password
This step helps to create a broker key along with encryption

Step 4: Create a broker certificate by requesting the broker key which is created in step 3
This step uses the IP address or domain name to create a broker certificate

Step 5: Use the CA certificate to sign the broker certificate which is created in step 4
This step uses the CA certificate along with a verified key and signature

Step 6: Accumulate the following file CA key, CA certificate, Broker key, and Broker certificate files and store them in a single folder

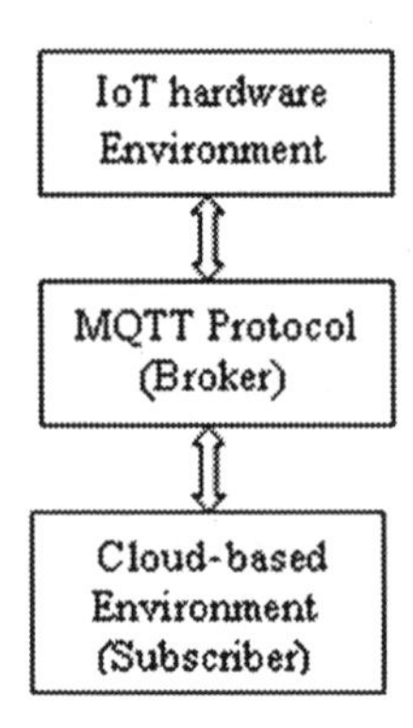

FIGURE 5.9
Secure MQTT protocol stages.

Step 7: Copy the CA certificate to the client
This step helps to copy the certificate to the client for the server process

Step 8: Edit the Mosquitto broker configuration file
This step helps to edit a particular broker configuration file

Step 9: Edit the client script file to use TLS and CA certificate
The Publisher and subscriber script come along with mosquitto broker as default code.

5.5.6 MQTT Protocol with Cloud Environment

In this module, classified EEG signals are communicated with Cloud Environment through a secure MQTT protocol for better data storage, processing, and analysis. This module flow diagram is given below in Figure 5.9. It has the main advantage of fast data transfer with a small overhead in IoT-based applications. IoT information can be stored or shared with suitable private or public brokers. Secure MQTT is also one of the popular brokers to support cloud services and makes it easy to integrate with third-party applications.

5.6 Conclusion

The Industry 4.0 revolution promises secure healthcare technology, which is one of the upcoming domains with the integration of biomedical information and internet connection. This smart healthcare system is more efficient and comfortable for the people who find it difficult to reach hospital. This chapter mainly focuses on EEG singles along with the IoT environment. This chapter also discussed the usage of EEG sensors, IoT hardware platforms, and IoT protocols. Received EEG signals were classified through the SVM machine learning algorithm and transferred the bioinformation in the MQTT protocol. This chapter also gave an insight view of various security issues related to Industry 4.0 secure IoT protocol. Usage of secure MQTT in terms of providing a secure username, password, TLS/SSL certification, and Arduino IoT platform was elaborated in this chapter.

References

1. Panwar, A., Malhotra, N., Malhotra, D., *INDUSTRY 4.0: A comprehensive review of artificial intelligence, machine learning, big data and IoT in psychiatric health care. Proceedings of 3rd International Conference on Computing Informatics and Networks*, Page 495–504, 2021.
2. Mashal, I., Alsaryrah, O., Chung, T.Y., Yang, C.Z., Kuo, W.H., Agrawal, D.P., Choices for interaction with things on internet and underlying issues. *Ad Hock Network*, 28, Page 68–90, 2015.
3. Said, O., Masud, M., Towards Internet of things: Survey and future vision. *International Journal of Computer Network*, 5, Page 1–17, 2013.
4. Sanzgiri, A., Dasgupta, D., *Classification of insider threat detection techniques. Proceedings of the 11th Annual Cyber and Information Security Research Conference*, pp. 25, 2016.
5. Business Logic Attack. n.d. http://whatis.techtarget.com/definition/business-logic-attack
6. Islam, S.R., Kwak, D., Kabir, M.H., Hossain, M., Kwak, K.S., The internet of things for health care: A comprehensive survey. *IEEE Access*, 3, Page 678–708, 2015.
7. Jing, Q., Vasilakos, A.V., Wan, J., Lu, J., Qiu, D., Security of the internet of things: Perspectives and challenges. *Wireless Network*, 20, Page 2481–2501, 2014.
8. King, J., Awad, A.I., A distributed security mechanism for resource-constrained IoT devices. *Information*, 40, Page 133–143, 2016.
9. Babar, S., Stango, A., Prasad, N., Sen, J., Prasad, R., *Proposed embedded security framework for internet of things (IoT). Proceedings of the 2nd International Conference on Wireless Communication, Vehicular Technology, Information Theory and Aerospace & Electronics Systems Technology (Wireless VITAE)*, Page1–5, 2011.
10. Sundaram, B.V., Ramnath, M., Prasanth, M., Sundaram, V., *Encryption and hash based security in Internet of things. Proceedings of the 3rd International Conference on Signal Processing, Communication and Networking (ICSCN)*, Page1–6, 2015.
11. AlShuhaimi, F., Jose, M., Singh, A.V., *Software defined network as solution to overcome security challenges in IoT. International Conference on Reliability, Infocom Technologies and Optimization (Trends and Future Directions)*, Page491–496, 2016.

12. Oke, J.T., Agajo, J., Nuhu, B.K., Kolo, J.G., Ajao, L.A., Two layers trust-based intrusion prevention system for wireless sensor networks. *Advanced Electronics & Telecommunication Engineering*, 1, Page 23–29, 2018.
13. Gyory, N., Chuah, M., *IoTOne: Integrated platform for heterogeneous IoT devices. International Conference on Computing, Networking and Communications (ICNC)*, Page 783–787, 2017.
14. Radanliev, P., De Roure, D., Nicolescu, R., Huth, M., Santos, O., Artificial intelligence and the internet of things in Industry 4.0. *CCF Transactions on Pervasive Computing and Interaction*, 3, 329–338, 2021.
15. Yan, Z., Zhang, P., Vasilakos, A.V., A survey on trust management for internet of things. *Journal of Network Computer Application*, 42, Page 120–134, 2014.
16. Roman, R., Zhou, J., Lopez, J., On the features and challenges of security and privacy in distributed internet of things. *Computer Network*, 57, Page 2266–2279, 2013.
17. Stankovic, J.A. Research directions for the internet of things. *IEEE Internet Things*, 14, Page 3–9, 2014.
18. Borgohain, T., Kumar, U., Sanyal, S., Survey of security and privacy issues of Internet of Things. arXiv preprint arXiv:1501.02211, 2015.
19. Sun, W., Cai, Z., Liu, F., et al., *A survey of data mining technology on electronic medical records. Proceedings of the International Conference on E-Health Networking, Application and Services*, Page 1–6, 2017.
20. Hu, J.-X., Chen, C.-L., Fan, C.-L., Wang, K.-H., An intelligent and secure health monitoring scheme using IoT sensor based on cloud computing, *Journal of Sensors*, 2017, 2017.
21. Govaert, J. A., Bommel, A. C., Dijk, W. A., Leersum, N. J., et al, Reducing healthcare costs facilitated by surgical auditing: A systematic review. *World Journal of Surgery*, 39(7), Page1672–1680, 2015.
22. Sun, W., Cai, Z., Li, Y., et al., Security and privacy in the medical internet of things: A review. *Security and Communication Network*, 2018, 2018.
23. Alnemari, M.H., Integration of a low cost EEG headset with the internet of thing framework. *UC Irvine Electronic*, 2017.
24. Zainuddin, B.S., Hussain, Z., Isa, I. S., *Alpha and beta EEG brainwave signal classification technique: A conceptual study. IEEE 10th International Colloquium on Signal Processing & its Applications (CSPA2014)*, 2014.
25. Chen, C., Loh, E.-W., Kuo, K. N., Tam, K.-W., The times they are a-Changin' – Healthcare 4.0 is coming!, *Journal of Medical Systems*, 44, 40, 2020.
26. Ajmera, P., Jain, V., Modelling the barriers of health 4.0–the fourth healthcare industrial revolution in India by TISM. *Operations Management Research*, 12, 129–145, 2019.
27. Li, J., Carayon, P., Health Care 4.0: A vision for smart and connected health care. *IISE Transactions on Healthcare Systems Engineering*, 2472, 2021.
28. Aceto, G., Persico, V., Pescapé, A., Industry 4.0 and health: Internet of things, big data, and cloud computing for healthcare 4.0. *Journal of Industrial Information Integration*, 18, 100129, 2020.
29. Javaid, M., Haleem, A., Industry 4.0 applications in medical field: A brief review. *Current Medicine Research and Practice*, 9, 102–109, 2019.

30. Haripriya, A. P., Kanagasabai, K., Secure-MQTT: An efficient fuzzy logic-based approach to detect DoS attack in MQTT protocol for internet of things. *Journal of Wireless Computer Network*, 90, 2019.
31. Patel, C., Doshi, N., A Novel MQTT security framework in generic IoTmodel. *Procedia Computer Science*, 2020.
32. Sochor, H., Ferrarotti, F., Ramler, R., An architecture for automated security test case generation for MQTT systems. *Database and Expert Systems Applications*, 1285, 2020.
33. https://www.wikiwand.com/en/10%E2%80%9320_system_(EEG)
34. Arduino official website: www.arduino.cc
35. Raspberry official website: www.raspberrypi.org
36. EEG Dataset: https://sccn.ucsd.edu/~arno/fam2data/publicly_available_EEG_data.html
37. MQTT official website: https://www.mqtt.org/
38. HivemqBroker official website: http://www.broker.hivemq.com
39. MosquittoBroker official website: https://www.test.mosquitto.org
40. IoT Eclipse Broker official website: http://www.iot.eclipse.org
41. Openssl official website: http://slproweb.com/
42. ftp://sources.redhat.com/
43. http://www.steves-internet-guide.com/mosquitto-tls/

6

Cybersecurity for Critical Energy Protection in Smart Cities

S. Karthi, N. Narmatha and M. Nandhini
V.S.B. Engineering College, Karur, India

CONTENTS

DOI: 10.1201/9781003203087-6

6.1 Introduction

Cybersecurity is the demonstration of securing PCs, laborers, cell phones, electronic structures, associations, and data from malignant assaults. It's otherwise calleddata innovation security or electronic data security. In the information society, cybersecurity has as of late become a significant idea in human existence, contingent upon the improvement of data and media transmission advances. The beginning of "cyber" emerges from the beginning of "cybernetic" which is a highlighted idea in characterizing the domination and correspondence of creature and instrument frameworks. "Cyber Space" is an extra idea that has a similar beginning and draws the limits of cybersecurity's calculated explanation [1]. There are six key ideas for better comprehension of network protection's degrees: privacy, uprightness, accessibility, validation, approval, and non-repudiation. Cybersecurity has arisen as a reflection and needs concerning these movementsand advancements. A smart city can be accepted as an additional impression of the data lifetime and civilization. Smart cities are covered by cutting-edge gear and programming of most recent innovations like modern control frameworks, the Internet of Things (IoT), Administrative Dominance and Information Accession and Dispersed Control Systems, and basic foundations interconnected with these fragments. Contingent upon this cyber organization among objects, constructions, and programming, past being an actual design, fundamental systems have become digital - genuine developments in shrewd urban communities. In the present profoundly interconnected world, solid energy conveyance requires advanced intense energy movementframeworks. Indeed, the country's security, monetary thriving, and the prosperity of our residents rely upon a solid energy framework. In that capacity, a first concern for the Office of Cybersecurity, Energy Security, and Emergency Response (CESER) is to make the country's charged force matrix and oil and flammable gas framework strong to cyber dangers. Today, 55% of the total populace lives in metropolitan regions, yet, it is normal that by 2050, practically 70% of the world be essential for metropolitan urban areas. This has raised the stress over how much stress to overpopulated and gridlock to even security stresses with rising interest for utilities. Cities consume75% of the world's normal assets, including 80% of the worldwide energy supply, and produce roughly 75% of the overall fossil fuel byproducts, as per reports. This has made improving energy utilization a central goal of a keen city. Scientists at the University of California Los Angeles (UCLA) have created a straightforward solar panel that can be mounted on the windows of structures to catch more daylight than a customary roof-mounted panel. Another advancement is a little, ultra-light wind turbine incorporated into a structure or other metropolitan construction. These areas are now being used or going through preliminaries throughout the planet, from the Eiffel Tower to Bahrain's World Trade Center and the Pearl River Tower in Guangzhou, China. The up-and-coming age of brilliant smart cities will profit with inventive approaches to coordinate renewable energy and energy-efficient and smart structure advancements.

6.2 Cybersecurity

Cybersecurity alludes to the assortment of advances, cycles, and practices expected to ensure associations, devices, activities, and data from assault, damage, or smuggledingress. Cybersecurity may likewise be alluded to as data innovation security. Cyberattacks are mainly aimed at accessing and destroying delicate data. Cybersecurity protects not only our software but also hardware that stores, handles, and transmits the data. Cybersecurity is the security of web-related structures like gear, programming, and information from cyber dangers. The preparation is utilized by individuals and challenges notwithstanding unapproved authorization to worker ranches and further automated structures [2]. Cybersecurity is useful in different fields such as application security, data misfortune counteraction, forensics, incident reaction, network security, security design, threat insight, vulnerability

FIGURE 6.1
CyberSecurity.

of the executives, and in the energy sector, smart city communities, and so on. User Authentication is an example of cybersecurity (Figure 6.1).

6.2.1 Advantages of Overseeing Cybersecurity

- Shield organizations anddata from unapproved access.
- Further developed information security and business movement on the board.
- Further developed accomplice trust in your information security courses of action.
- Further developed association capabilities with the right security controls set up.
- Quicker recovery times in the event of a break.

6.3 Cybersecurity in Smart Cities

A smart city is a metropolitan area that utilizes different kinds of electronic techniques and actuators to accumulate data. Pieces of information procured from that data are used to administer assets, resources, and organizations beneficially; subsequently, that data is used to work on the tasks across the city. As smart urban communities become more interconnected, and the degree of computerized framework turns out to be more intricate, these administrations will turn out to be more defenseless against cyberattacks [3]. Indeed, even the littlest weakness can be abused to incredible impact, and smart urban communities are just pretty much as solid as their most vulnerable connection. Smart cities are endangeredbycyber assaults from numerous points of view. [4] Advanced Persistent Threats (APTs) are probably the most hazardous. These dangers depend on a few unique assaults working as one to disturb metropolitan administrations, frequently utilizing malware and "zero-day" programming weaknesses. IoT advances are especially powerless, and keeping in mind that it's feasible to fix any uncovered region, programmers can do enduring harm.

A portion of the cyberattacks that cloud harm smart city framework is as follows [4].

6.3.1 Resource, Information, and Identity Threat

Information robbery is seemingly the most notable digital wrongdoing. Programmers can invade information banks and take eventually conspicuous information. Smart city framework is especially defenseless against this, and programmers have been known to extricate individual information from public installment foundations with genuine outcomes.

6.3.2 Man-in-the-Middle Assaults

A site is a point at which a hacker can intrude on correspondence between two gadgets and pose as the sender, sending bogus data to raise a ruckus. For instance, a programmer might get to a portability stage and report public vehicle delays, which could prompt more individuals to take a vehicle to work, causing a convergence in rush hour gridlock that carries a city to a halt.

6.3.3 Hijacking Gadgets

Gadget hijacking is one of the additional alarming parts of cybercrime. Utilizing security weaknesses, assailants can assume responsibility for a gadget and make use of it to upset a cycle. Traffic signals and street signals are especially defenseless.

6.3.4 Physical Disruption

Older style actual power can likewise be utilized to think twice about the complexities associated with the network. As numerous frameworks depend on complex cycles and input from organizations of sensors, actual harm to any part could cause a chainresponse of harm.

6.3.5 Ransomware

The entirety of the above could be utilized to clutch a city to deliver. Programmers, or pirates, utilize these to think twice about the interaction or delivery of secret information except if certain requests are met. Paying a payoff would start a perilous trend.

6.3.6 Disseminated Denial of Service

Disseminated Denial of Service (DDoS) assaults are straightforward. A pirate can overpower a framework by assaulting it with demands, obstructing the assistance of the individuals who need it. With genuine clients incapable to get help, city frameworks will neglect to help their residents.

Cyber dangers to smart urban areas should be viewed extremely in a serious way [5].

Potential plans incorporate:

1. Formation and utilization of safety inspect records for encryption, verification, authorization, and programming refreshes while carrying out new systems.
2. Execution of safeguards andinstructionsabrogates on all urban frameworks.
3. Improvement of activity ideas and methods for reacting to cyber assaults.

6.4 Key Cybersecurity Steps

There are five key cybersecurity steps that pioneers take undeniably more frequently than different urban communities to address their cybersecurity weaknesses [6]:

6.4.1 Give Cybersecurity Training to Staff

This is a basic advance for urban communities since cybercriminals frequently exploit workers' missteps.

6.4.2 Focus on Resources and Make Access Control Arrangements

Securing a city's most significant resources is a keen initial step, as is ensuring the city forces tight controls on who can get to its frameworks.

6.4.3 Create a Cyber Episode Reaction and Recuperation Plan

Smart city pioneers comprehend they should not just act rapidly to stop an assault, but, in addition, should have measures set up to restrict the eventual outcomes, including those identified with obligation and monetary and reputational impacts.

6.4.4 Put Resources into Calamity Recuperation, Reaction, and Occasion the Executive's Innovation

Regardless of how solid a city's firewalls are, it just takes one miscreant to traverse. So brilliant city pioneers put all the more vigorously in specific recuperation and reaction innovation to act rapidly to relieve impacts.

6.4.5 Secure Basic Framework

This incorporates security testing of power lattices, traffic signals, medical clinics, and other metropolitan resources. Interconnecting city resources and areas through IoT and different innovations can open urban communities to a cataclysmic assault on the off chance that they don't satisfactorily defend their foundation.

Indeed, one indication of a smart city pioneer that is generally exceptional in utilizing innovation and inventive arrangements is its degree of cybersecurity. Ninety-five percent (95%) of metropolitan networks assigned pioneers in the examination said they were all set for cyberattacks, against only 8% of fledgling cities [6] (Figure 6.2).

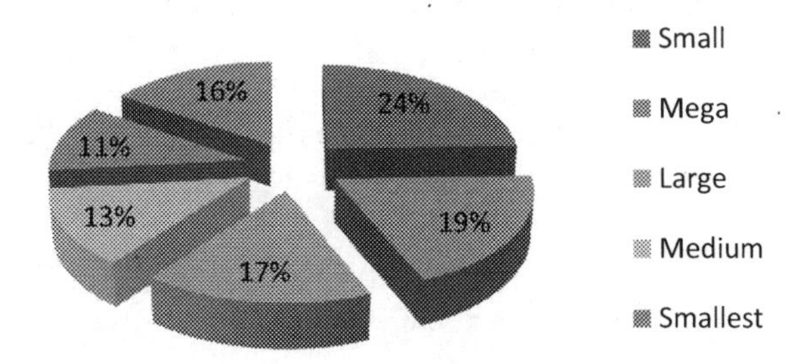

FIGURE 6.2
Percentage of cities ready for cyberattacks.

6.5 Energy Production in Smart Cities

The smart city is aviable and useful metropolitan local area that gives a better grade of life than its tenants with an optimal organization of its resources, where clean and smart energy age is a central question. Under this setting, dispersed age can furnish a sufficient device to manage energy of unwavering quality and to effectively carry out sustainable sources; by and by, choice and scaling of energy frameworks, thinking about area, is not an inconsequential undertaking [7]. Smart cities are depended upon to end up being more freeand deal with their energy impression all the more productively, considering nearby assets and the necessities of different partners. In this unique circumstance, Smart Energy Management (SEM) involves understanding the capability of energy as a fundamental structure block for smart cities [8]. The smart energy frameworks can lessen the essential energy utilization of cities and cover the leftover interest by exploiting the nearby environmentally friendly power assets, contingent upon the climatic qualities of every city and the productivity of the energy vectors [9].

Proper energy management requires precise metering. Multi-work, conveying smart meters that estimate energy traded and imported, request and force quality, and the board of burden, nearby age, client data, and other value-added capacities are fundamental when making smart networks to arrange market interest. Another key segment is the utilization of smart energy sensors with various capacities to gather and share information for prescient examination. This information can be utilized to identify and foresee energy needs and give important experiences during seasons of pinnacle interest. The up-and-coming age of smart cities will profit with inventive approaches to incorporate environmentally friendly power and energy-efficient and keen structure innovations [1]. The IoT is the organization of interconnected items or gadgets installed with sensors and cell phones which can produce information and convey and share that information. As self-learning structures become more far and wide, innovatively progressed structures will want to discuss electronically with one another to guarantee that energy utilization is adjusted. A significant element of a smart city is the examination and utilization of information gathered by IoT gadgets and sensors to further develop the foundation, public utilities, and administrations, just as for prescient investigation. As the IoT grows, so does the requirement for powerful cybersecurity protection against cyberattacks on IoT-associated gadgets, applications, and organizations [10] (Figure 6.3).

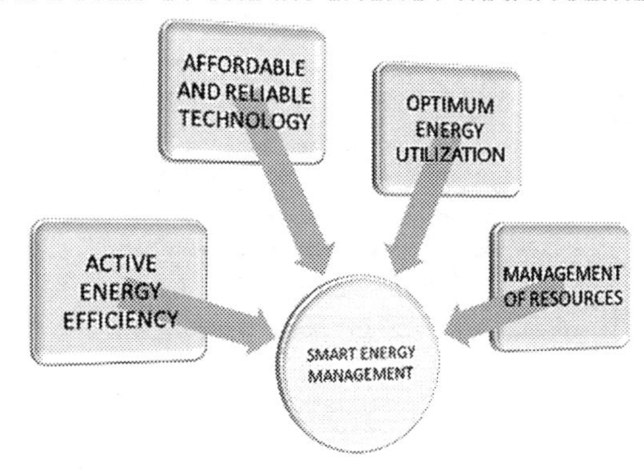

FIGURE 6.3
Features of Smart Energy Management (SEM).

6.6 Microgrids

A microgrid is a decentralized gathering of power sources and loads that regularly work associated with and simultaneous with the conventional wide-region coordinated framework. A microgrid can adequately coordinate different wellsprings of dispersed age (DG), particularly sustainable power sources. Another age of low carbon microgrids is changing the manners by which thickly populated cities plan and work utility frameworks utilizing the idea of privately created and burned-through energy. Microgrids permit prescient upkeep and are especially encouraging for guaranteeing strength in the energy requests of cities.

The smart microgrid's cyber-physical model contains four layers [11]:

- Physical power system layer
- Sensor and actuator layer
- Communication layer
- Management and control layer (Figure 6.4)

6.6.1 Physical Layer

The physical layer contains the microgrid's power parts, like transformers, dynamos, power devices, converters, switches, and weights.

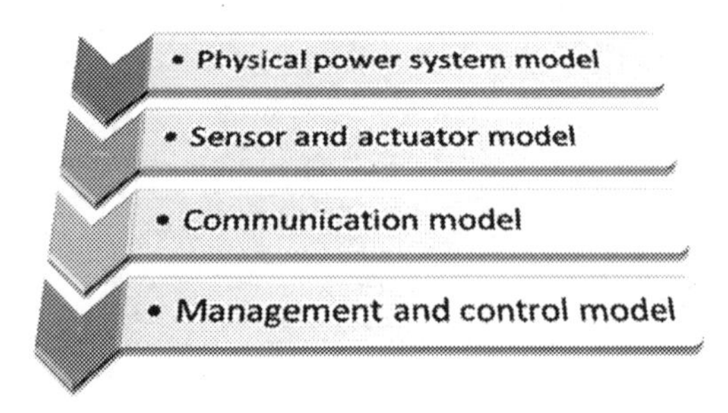

FIGURE 6.4
Smart microgrid's cyber-physical model.

6.6.2 Sensor and Actuator Layer

The sensor and actuator layer comprises actuators, estimation gadgets, and gadgets to carry out the control choices (built in the management layer). The actuators and estimation gadgets are answerable for estimating data about the framework's condition, together with potential, recurrence, voltage, and electrical switch level. The actuators and power gadgets incorporate generator regulators, circulated age regulators, and transfers of circuit breakers.

6.6.3 Communication Layer

The communication layer comprises gadgets like routers, circuit breakers, and the communication channel and is answerable for data trade among pertinent levels. In smart microgrids, the connection framework can be cabled or remote, contingent upon framework necessities.

6.6.4 Management and Control Layer

The administration layer is a focal domination framework that is in risk for microgrid action underneath different states [12]. This layer gets appraisal layer information through the correspondence layer and passes on control signals for the sharp smaller than usual cross sections' ideal activity. The control signals are yet again delivered off sensors between the correspondence layer.

6.7 Cyberattacks on the Energy Sector

Cyber assaults are on the riseand the energy area is a significant objective for hoodlums. Energy frameworks have transformed into profoundly conveyed frameworks which require proactive insurance. Following are the three factors that expand the weakness of the energy area [13].

- The fast rateof technological advancement
- The expanding refinement of cyber assaults
- The area's allure as a cyber target

6.7.1 The Fast RateofTechnological Advancement

The difference in the energy structure is currently in progress. It's driven by the necessities stretch out permission to get energy, and empowered by arising imaginative specialized arrangements. In the most fundamental sense, mechanical development supposedly is a basic empowering agent of progress. Be that as it may, the remarkable development in mechanical advancements in the energy area assembles the degreesof unpredictable and focused cyber assaults. Because of that, energy organizations are forcefully creating procedures to battle their security inadequacies and lift the security of their advanced resources.

6.7.2 The Expanding Refinement of Cyber Assaults

It is for the most part acknowledged that cyber assaults have becomemore complex throughout the long term. These kinds of assaults may come from coordinated wrongdoing gatherings, mechanical reconnaissance groups, cyber fear mongers, and even country states. In addition, these multi-vector assaults abuse obscure and complex weaknesses, causing essentially adverse consequences for an enormous scope [14]. In this way, the rising refinement of cyber assaults in the energy area has the exceptionally potential to harm enormous quantities of substances across huge geographic districts.

6.7.3 The Area's Allure as a Cyber Target

The energy industry is an IP-escalated industry. At the end of the day, it holds monstrous licensed innovation. It is an obvious fact that IP is at the center of the seriousness of numerous associations. For the most part consequently, it's an alluring objective for cyber hoodlums and cyber undercover work. Cyber reconnaissance against the energy area might be established with political and monetary intentions, which may give the entertainer admittance to information that presents an innovative benefit, comprising a likely danger to the energy security [7].

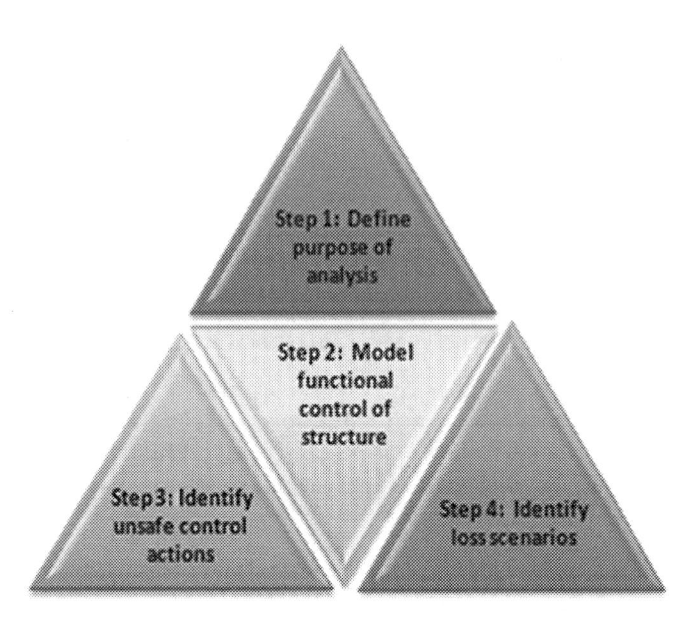

FIGURE 6.5
Steps for protecting our energy from cyberattacks.

6.8 Steps for Protecting Our Energy from Cyberattacks

There are four steps for protecting our energy from cyberattacks. Figure 6.5 shows the steps for protecting our energy from cyberattacks. They are

Step 1: Define the motivation behind the examination
Step 2: Model practical control structure
Step 3: Identify hazardous control activities
Step 4: Identify misfortune situations

6.9 Renewable Energy in Smart Cities and Cybersecurity

Cities' small geological impression gives a false representation of their importance. They enfold 2% of the world's mainland, yet represent a large portion of the total populace, financial exercises, and power use. Here, we center on the third perspective – power consumption – as urban areas and inexhaustible power have, separately, become the regular environmental elements and force of choice all around. The two are progressively indivisible [15].

As urban communities strive to draw in developing organizations, ability, and advancement in an undeniably worldwide contest, smart sun- and wind-based energy have becomethe lead for some in accomplishing their smart urban community objectives [16]. Smart metropolitan networks concentrates with care in conveying two fuel sources that line up with their objectives: solar and wind. benefits is accepting wind and sunlight-based force as they arrive at cost and execution equality with traditional fuel origin across the world, assist to cost-adequately stabilize the network, and become more important resources on account of progressively savvy stockpiling and other new advancements (Global renewable energy trends) [16] (Figure 6.6).

These renewable energy sources presently come nearest to satisfying the developing need for dependable, moderate, and naturally capable fuel sources that utilities try to give. Thus, renewable energy has become the favored fuel hotspot for key customers like urban communities [17]. Cybersecurity is the security of interconnected electric force frameworks againstcomputerized assaults. Solar is one of the various electric age headways used on the cross section, adding to enormous scope age as solar farms and utility-scale establishment, similarly as restricted scale scattered energy resource age in the construction charitable rooftop establishment, storage frameworks, and microgrids. Renewable energy cybersecurity assaults can happen in a heap of ways. For instance, phishing messages might be focused on clueless staff with astutely masked pernicious connections containing malware or ransomware. A denial-of-service assault may see a culprit hack into a framework and square staff access while taking control themselves. Bogus information can be infused into data streams, making a bogus story of occasions bringing about awful choices. Also, "cascading" can be a

FIGURE 6.6
Cybersecurity and renewable energy in smart cities.

major issue whereby one framework is contaminated and rapidly goes to an optional and reinforcement one, making a colossal extension blackoutthat is difficult to switch.

6.10 Challenges in Cybersecurity

6.10.1 Ransomware

As stated by Cybersecurity firm Sophos, nearly 82% of Indian affiliations were thrashed by ransomware in the latest half-year. Ransomware assaults are fundamental for particular customers, yet more so for associations who can't get to the data for managing their day-by-day activities. Notwithstanding, with most ransomware assaults, the assailants don't deliver the information although the paymentis made and rather attempt to coerce more cash [18].

6.10.2 IoT Attacks

In consonance withIoT Analytics, almost 11.6 billion IoT devices are up there by 2021. As the apportionment of IoT contraptions is growing at a remarkable rate, so are the hardships of Cybersecurity. Assaulting IoT devices can accomplish the accommodate of sensitive client information. Guaranteeing IoT contraptions is undoubtedly the best testin Cybersecurity, as getting to these contraptions can unlock the entrances for other pernicious assaults (Figure 6.7).

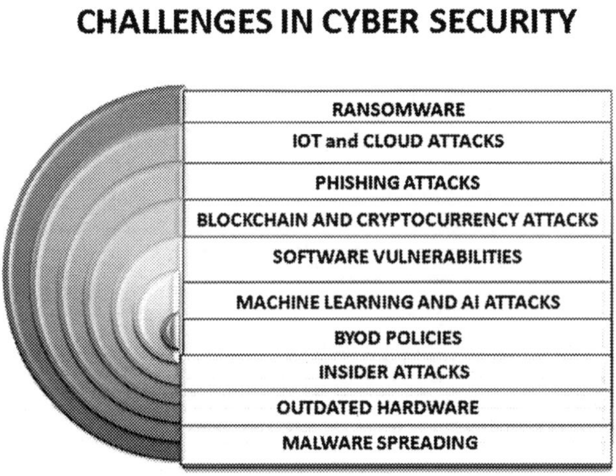

FIGURE 6.7
Challenges of Cybersecurity.

6.10.3 Cloud Attacks

Generally, nowadays, almost everyone uses cloud organizations for individual and master necessities. Additionally, hacking cloudstages to take customer details is one of the troubles in Cybersecurity for associations.

6.10.4 Phishing Attacks

Phishing could be the initial step of putting clients and associations in danger, as it is exceptionally simple to execute the cycle [19]. Phishing is a kind of well-disposed planning attack mostly used to take customer data, including login capabilities and charge card numbers. Unlike ransomware attacks, the developer, in the wake of getting to ordered customer data, doesn't obstruct it. Taking everything into account, they use it for their advantages, for instance, web-based shopping and illicit money moving.

6.10.5 Blockchain and Cryptocurrency Attacks

While blockchain and cryptocurrency probably won't mean a lot to the normal web client, these innovations are an immense arrangement for organizations. Accordingly, assaults on these systems present extensive difficulties in Cybersecurity for organizations as they can think twice about client information and business tasks.

6.10.6 Software Vulnerabilities

Refreshing your gadget's product with the farthest down the line structureought to be the main concern. A more established programming rendition may hold blotches for security weaknesses that are made stable by the engineers in the more current variant. Assaults on unpatched programming adaptations are one of the critical troubles of Cybersecurity. These attacks ordinarily bring off a colossal quantity of individuals, like the Windows zero-day attacks.

6.10.7 Machine Learning and AI Attacks

While Machine Learning and Artificial Intelligence advances have demonstrated profoundly advantageous for gigantic improvement in different areas, they havetheir weaknesses also. These innovations can be abused by illegal people to do cyberattacks and present dangers to organizations. These innovations can be utilized to perceive excessive regard centers amonga huge datafile.

6.10.8 BYOD Policies

Many associations have a Bring-Your-Own-Device (BYOD) strategy for their representatives. Having such frameworks represents various difficulties in

Cybersecurity. Initially, if the gadget is managing an obsolete or pilfered variant of the product, it is as of now, a brilliant vehicle for hackers to get to. Since the technique is being utilized for individual and expert causes, hackers can, without much stretch, ingress private business information. Furthermore, these gadgets make it simpler to get to your private organization if their security is compromised.

6.10.9 Insider Attacks

While most difficulties of Cybersecurity are outer for organizations, there can be cases of an inmost job. Representatives with avindictive aim can release or fare private information to contenders or others. This can provoke colossal money-related and reputational disasters for the business. Presenting firewall gadgets for controlling data along a concentrated specialist or confining permission to reports reliant upon maintaining sources of income can help with restricting the prospect of insider attacks.

6.10.10 Outdated Hardware

All things considered, don't be astounded. Not all difficulties of Cybersecurity come as programming assaults. With programming engineers understanding the danger of programming weaknesses, they offer an occasional update. Notwithstanding, these new updates probably won't be viable with the equipment of the gadget. This is the thing that prompts obsolete equipment, wherein the equipment isn't progressed adequately to manage the most recent programming adaptations. This leaves such gadgets on a more established rendition of the product, making them exceptionally powerless againstcyberattacks [18].

6.10.11 Malware Spreading

The critical peril the Smart Grid can be looked at by Malware spreading which is a significant concern. The aggressors can create malware that can be skewered into contaminating the association workers just as tainting the gadgets. By utilizing the malware spreading, the aggressor can control the elements of gadgets or the frameworks which will permit the assailants to get entrance for gathering touchy data [19].

6.11 Challenges of Cybersecurity in Smart Cities

This is a test,it is a chance to propel the aggregate advancements toward the fulfillment of solar cybersecurity around the world [20]. Brilliant urban areas are progressively enduring an onslaught by an assortment of dangers. These incorporate modern cyberattacks on a basic framework, bringing industrial control systems (ICSs) to a pounding end, mishandling low-power wide-area networks (LPWAN) and gadget correspondence seizing, framework lockdown dangers brought about by ransomware, control of sensor information to cause boundless frenzy (e.g., fiasco location frameworks) and siphoning resident, medical care, shopper information, and personally identifiable information (PII), among numerous others, explains DimitriosPavlakis, Industry Analyst at ABI Research. In this undeniably associated mechanical scene, each keen city administration is pretty much only as secure as its most vulnerable connection. In the recent past, smart cities were just conceptualized in science fiction films and books of fiction [1] (Figure 6.8).

Today, in any case, that thought is quickly changing from inventive domains into real factors. Essentially, as smart city networks spring up across the globe, they likewise have made an intriguing security-risk perspective. Subsequently, this is tortured by cybercriminals who make plans to cause ruin at the littlest opportunity. Security susceptibilities in the Smart City Foundation, Complex and Massive Attack Area, Ramifications of a Cyberattack, Obscured Oversight and Arranging, Budgets and Political Shifts are some of the top difficulties of cybersecurity in smart cities [21]. To challenge environmental change, we will add increasingly more feasible creations to the lattice, so a greater amount of inverters (and perhaps at the

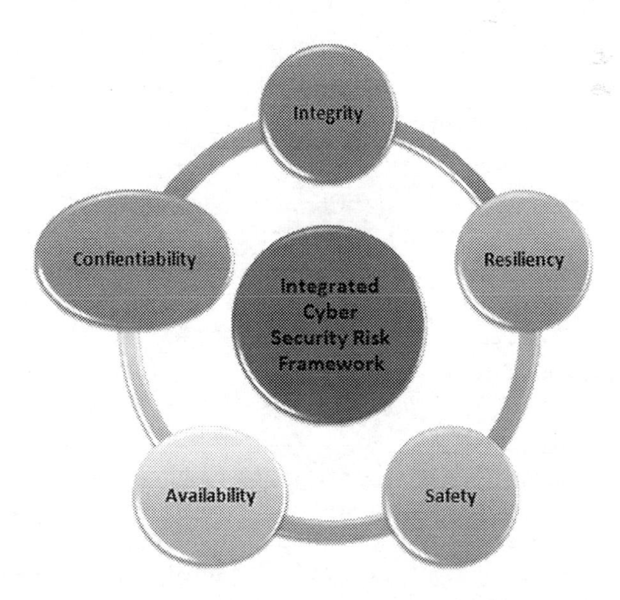

FIGURE 6.8
Integrated cybersecurity risk framework.

same time, batteries) will be associated. At the same time, we're progressively changing utilization from petroleum derivatives for warming to power or from interior ignition motors to electric ones.

6.12 Challenges of Cybersecurity in Renewable Energy

The cybersecurity danger is genuine and surprisingly the greatest and solar establishments aren't insusceptible to it. The digitalization of force matrices with expanding the availability of solar power plants put the PV resources in high danger by ill-conceived programmers who can break into the network security, which can put the force supply in danger. Solar energy can't be put away endlessly, and delivering new force takes up a ton of significant time [20]. Cybersecurity occasions on the network are of critical concern, given the expanded organization of savvy lattice advances and different types of wise controls and ICSs. Over the most recent couple of months, a few remarkable clean energy organizations have found ways to diminish the danger of a break [22]. The solar energy sector is honing its attention on solar cybersecurity safety in the midst of developing worries about digital dangers.

Some of the challenges of cybersecurity in wind energy are:

- Increasing cyber occurrences focusing on wind energy frameworks
- Difficultyin set up compelling cybersecurity rehearses
- Wind energy resources require powerful cybersecurity rehearses
- No single cybersecurity methodology can apply to all wind plants
- Available cybersecurity alternatives might be excessively expensive
- Few breeze explicit cybersecurity principles exist
- Few motivating forces to focus on cybersecurity [23].

6.13 Conclusion

In the present world, Cybersecurity is a pivotal piece of any business. Cybersecurity is as yet a moderately new piece of IT and has changed fundamentally actually as it has been seen as an alternate discipline inside IT security. Executing solid protections against cyber assaults expects admittance to a gifted, experienced cybersecurity labor force. As digital perils are growing over time one day to another, cybersecurity is fundamental in the field of energy creation in smart urban communities. Both the number and security ramifications of modern cyber assaults on organizations giving basic energy frameworks are expanding. As force organizations and, partially, oil and gas frameworks both upstream and downstream are turning out to be progressively incorporated with data correspondence innovation frameworks, they are developing as more powerless to cyber assaults. Therefore, cybersecurity significantly affects the present world. Cybersecurity, key steps, energy production in smart cities, microgrids with cybersecurity, and cyberattacks on the energy sectorare discussed. After that, protecting steps and challenges are provided.

References

1. Göçoğlu, V. (2019). Cyber security of critical infrastructures in smart cities. *Uluslararası Yönetim Akademisi Dergisi*, 2(1), 51–63.
2. Graham, J., Olson, R., & Howard, R. (Eds.). (2016). *Cyber security essentials*. Boca Raton, FL: CRC Press.
3. https://www.bsigroup.com/en-GB/Cyber-Security/
4. https://hub.beesmart.city/en/strategy/the-importance-of-cyber-security-and-data-protection-for-smart-cities
5. Alibasic, A., Al Junaibi, R., Aung, Z., Woon, W. L., & Omar, M. A. (2016, September). *Cybersecurity for smart cities: A brief review*. In *International Workshop on Data Analytics for Renewable Energy Integration* (pp. 22–30). Springer, Cham.
6. https://econsultsolutions.com/cybersecurity-a-smart-city-imperative/
7. Calvillo, C. F., Sánchez, A., & Villar, J. (2013, October). *Distributed energy generation in smart cities*. In *2013 International Conference on Renewable Energy Research and Applications (ICRERA)* (pp. 161–166). IEEE.
8. Barrionuevo, J. M., Berrone, P., & Ricart, J. E. (2012). Smart cities, sustainable progress. *IESE Insight*, 14(14), 50–57.
9. Wang, Y., & Yan, G. (2019). A new model approach of electrical cyber physical systems considering cyber security. *IEEJ Transactions on Electrical and Electronic Engineering*, 14(2), 201–213.
10. Prochazkova, D., & Prochazka, J. (2018, May). *Smart cities and critical infrastructure*. In *2018 Smart City Symposium Prague (SCSP)* (pp. 1–6). IEEE.
11. Nejabatkhah, F., Li, Y. W., Liang, H., & Reza Ahrabi, R. (2021). Cyber-security of smart microgrids: A survey. *Energies*, 14(1), 27.
12. Risk, C. P. I. (2015). The future of smart cities: Cyber-physical infrastructure risk.
13. https://www.swisscyberforum.com/blog/all-you-need-to-know-about-cyber-security-threats-in-energy-sector/
14. Morag, N. (2014). *Cybercrime, Cyberespionage, and Ccybersabotage: Understanding Emerging Threats*. Colorado Technical University: College of Security Studies. Available online at: https://www.coloradotech.edu/media/default/CTU/documents/resources/cybercrime-white-paper.pdf (accessed August 11, 2019)

15. Baig, Z. A., Szewczyk, P., Valli, C., Rabadia, P., Hannay, P., Chernyshev, M., ... & Peacock, M. (2017). Future challenges for smart cities: Cyber-security and digital forensics. *Digital Investigation*, 22, 3–13.
16. https://www2.deloitte.com/us/en/insights/industry/power-and-utilities/smart-renewable-cities-wind-solar.html
17. Pilipczuk, O. (2020). Sustainable smart cities and energy management: The labor market perspective. *Energies*, 13(22), 6084.
18. https://www.jigsawacademy.com/blogs/cyber-security/challenges-of-cyber-security/
19. https://www.aimspress.com/article/doi/10.3934/electreng.2021002?viewType=HTML
20. https://www.saurenergy.com/solar-energy-articles/cyber-security-is-the-new-challenge-for-renewable-energy
21. https://resources.infosecinstitute.com/topic/top-cyber-security-challenges-smart-cities/
22. Al-Mohannadi, H., Mirza, Q., Namanya, A., Awan, I., Cullen, A., & Disso, J. (2016, August). *Cyber-attack modeling analysis techniques: An overview*. In *2016 IEEE 4th international conference on future internet of things and cloud workshops (FiCloudW)* (pp. 69–76). IEEE.
23. https://www.researchgate.net/publication/338924438_Cybersecurity_for_Wind_Energy

7

Emerging Industry Revolution IR 4.0 Issues and Challenges

Azeem Khan
ADP Taylor's University, Selangor, Malaysia

Noor Zaman Jhanjhi
SoCIT Taylor's University, Selangor, Malaysia

R. Sujatha
SITE, Vellore Institute of Technology, Vellore, Tamil Nadu, India

CONTENTS

7.1 Introduction

The industrial sector is envisioned as the place where goods and services are produced. Industries are the backbone of the economy of any nation. Thus, industrial growth is marked as the major contributor to the overall GDP of the nation. As it is evident from history, the first industrial revolution acronymized as IR 1.0, revolutionized manual production toward machination with the discovery of steam power, and steam engines paved the way for the mechanization of the handloom industry, thereby increasing the

DOI: 10.1201/9781003203087-7

production with the implementation of machines running on steam power and this revolutionized the whole era; the industry which flourished with the advent of IR 1.0 was textile and steel industry and coal was the main energy source for this era. The second industrial revolution which is also acronymized as IR 2.0 started with the discovery of electricity, which introduced railways into the system, and electrical machines were introduced in the production of goods and services. Using this source of energy, factories increased their production on massive scales, and apart from coal, oil was also the main source of energy in this era, which boomed with the introduction of combustion engines, automobile industries enabled with assembly lines, and steel industries boomed a lot resulting in the production of trains for railways and motor-cars for automobile industry [1]. The third industrial revolution acronymized as IR 3.0 came into the limelight with the invention of electronics which led to the invention of computers and Information and Communication Technology (ICT); this paved the way toward the automation of many tasks resulting in the replacement of many tasks which were done by humans prior. This era is termed the era of digitization. From mechanization, electricity, and digitization, we have stepped into an era where the advent of the internet and its implementation in enabling communication with objects apart from communication among human beings and is termed the fourth industrial revolution acronymized as IR 4.0; it is conceived as a complex technological system comprising the Internet of Things (IoT), which is envisioned as the internet connection between physical objects that are part of our day-to-day life such as RFID sensors implanted with physical objects in shopping malls, retail groceries, in shipping, logistics that have the ability to interact with the system remotely: Cyber-Physical Systems (CPS), IoT, Industrial Internet of Things (IIoT), Robotics, Cloud Computing, Fog and Edge Computing, Big Data, Data Analytics which enabled visualization, 3D printing, and Augmented Reality. IR 4.0 is an emerging concept that is related to industries specifically the manufacturing sector wherein raw material gets into the system and a finished product is processed out as an output, and all these systems are interconnected through automated systems wherein minimal or no human intervention is required. The whole system is backed by Information and Communication Technologies (ICTs) which are connected with the internet by means of CPS incorporated through a combination of processing, transmission, and regulation of the entire objects connected through this system digitizing value chains across the organization, apart from conceiving product, product development, manufacturing, structuring, and providing services, it also entails internal operations of the businesses starting from suppliers to the customers' end [2,3]. This chapter's structure is as follows. The first section is an elaborate discussion on the introduction to I4.0. The second section is focused on a literature review highlighting the work done so far in I4.0. The third section involves the discussion on the concept of I4.0 and its inception. The fourth and fifth sections are focused on a discussion of the significance and relevance of I4.0 in the contemporary world. The sixth and seventh sections are focused on challenges and issues involved in I4.0. The last three sections are dedicated to the discussion on the findings from the works done by experts concerned with I4.0 and their reviews related to this field. Lastly, we have the conclusion which abridges the work of the whole findings available from the literature review and the future work, which is meant to throw light on the available future prospects with this new paradigm, which is basically a set of complex technologies advancing in the continuum.

7.2 Literature Review

Digitization of the manufacturing sector was started as Industrial Revolution 4.0 (IR 4.0) or, in brief, I4.0, back in November 2011 by the German Federal Ministry of Education and Research. I4.0 is conceived as a comprehensive concept introduced by Klaus Schwab, the executive chairman of the World Economic Forum; he envisioned I4.0 as the combination of hardware, software, and CPS with an emphasis on communication and connectivity encompassing Robotics, Artificial Intelligence (AI), nanotechnology, biotechnology, IoT, IIoT, 5G networking technology, 3D printing, and completely automated vehicles which connotes automation of manufacturing sector composed of intelligent autonomous ecosystems. The key design principles and goals of I4.0 include information transparency, technical assistance, and decentralized decision-making process through the interconnection between machines, devices, sensors, and people to connect and communicate via IoT and the Internet of People where the latter involves communication with the machines whereas the former involves humans communicating with each other. The whole idea of I4.0 was proposed as a high-tech strategy for the development of the industrial sector involving manufacturing in 2020. Later, other countries also started the same initiatives with different names such as "Made in China 2025" by china, "Social 5.0" by Japan, "Manufacturing USA" by the USA, etc., as stated in Table 7.1 [3–5].

Ever since 2011, globally, several governmental initiatives [6] have been taken by many countries under various names as shown in Table 7.1; the data provided here clearly state how the top ten developed world economies based on GDP as well are serious toward explicit commitments to adopt, guide, and implement ICT-based manufacturing sectors because these segment of industries have great potential to contribute toward the economy at the social level and for the stability of environment as well. In the USA, IR 4.0 is synonymous with smart manufacturing and the Internet Industry of Things. Succinctly, I4.0 is summed up as the digitalization and

TABLE 7.1

IR 4.0 Country-Wise Terminologies

Year	Country	Term
2011	Germany	Industrie 4.0
2013	Denmark	Manufacturing Academy of Denmark (MADE)
2014	Spain	Industria Conectada 4.0
2014	UK	Future of Manufacturing
2015	China	Made in China 2025
2015	Austria	Plattform Industrie 4.0
2016	Japan	Society 5.0
2016	Sweden	Smart Industry
2016	Italy	Piano Industria 4.0
2016	Belgium	Made Different
2017	Netherlands	Smart Industry
2017	USA	Manufacturing USA
2017	France	Industrie du Futur
2017	Czech Republic	Pr˙umysl 4.0
2017	Portugal	Ind´ustria 4.0

interconnection of all production units comprised in an economic system [7]. Apart from the involvement of technologies, I4.0 is also influencing various areas of enterprises inclusive of management, organization of work, regulatory frameworks, and toward the integration of enterprises as well. Integration of enterprises is anticipated in three aspects, namely horizontal, vertical, and end to end. To elaborate a bit, horizontal integration carries coordination between various processes involved across different value chains involved in an enterprise network. Vertical integration implies all-embracing automation within the enterprise and last of all end-to-end integration comprises digitization of all contributors, which includes communication between human and human or it may be between machine and machine or human and machine. The aforementioned scenario with the integration of the above three dimensions makes a whole enterprise network more dynamic beginning with adjustments in infrastructure, production systems, and, last but not least, the customers [5,8]. To make the whole automation of production processes possible, the communication between the objects connected in this plethora of things needs to be automated, this is possible with the intervention of the internet among things connected to this system, which is done with the help of IoT wherein, the objects can communicate with each other autonomously [9,10]. Three paradigms central to I4.0 are smart products, smart machines, and augmented operators. Smart products represent those products that can necessitate production resources and arrange all the production processes till the end of its completion. Smart machines represent CPS wherein, the whole production process is automated with a decentralized decision-making process, and also making processes self-organizing, adaptable, and coordinating between the networks involved in production. Supplemented operative is the human being who is going to leverage technology to implement workerless production services acknowledging the centrality of human presence to monitor, specify, and verify the production strategies [11]. I4.0, apart from automating the industrial sector involved in manufacturing, is also impacting other aspects of lives and has proven to be a useful asset to improve the production in the agricultural and food sector. The rise in global population is compelling the intervention of technology to provide smart services to automate the irrigation process, thereby increasing the productivity with available natural resources with limited availability of agricultural land due to diminishing natural resources. I4.0 can play a vital role in the agricultural sector by enhancing the operational efficiency resulting in the production of agricultural products, by implementing IoT technology that can help peasants in predicting the weather and availing them to make better decisions for enhanced productivity of their crops and also it facilitates by providing details in opting which crops to grow in a given circumstance [12].

The CPS which includes smart grids, autonomous automobiles, medical monitoring gadgets, industrial control systems, and robotics enactment is creating changes in the way humans work and in work organization, successively, establishing new challenges and avenues. To make the maximum out of available resources and manage the complications, there is a need to increase an all-encompassing comprehension of the rising socio-specialized associations and apply new human-driven methodologies and techniques to study the readiness of the industries toward embracing this new paradigm has been discussed by [13], highlighting the issues and challenges involved in embracing I4.0. Table 7.2 precisely mentions the prime features of industrial revolutions.

TABLE 7.2

Prime Features of Industrial Revolutions [1]

Time Span	Source of Energy	Inventions	Industries	Conveyance
2000–...	Green Energies	Internet, 3D Printer, Genetic Engineering	High Tech Industries	Electric Cars, Ultra-Fast Trains
1960–2000	Natural Gas, Nuclear Energy	Computers, Robots	Automobiles, Chemical	Car, Plane
1900–1960	Electricity, oil	Combustion Engines	Automobiles, Metallurgy	Train, Car
1760–1900	Coal	Steam Engine	Textile, Steel	Train

7.3 Industry 4.0 Concept and Its Significance

The significance of I4.0 can be affirmed categorically with the numerous advantages this paradigm provides in terms of positive outcomes it envisages, ranging from productivity, mass customization, flexibility, reducing time, and cost of the products till they are sent to market, improved quality of the goods and services, reduced lead time, high environmental sustainability, environment friendly, improved economic growth thereby contributing to the reduction of unemployment and creating better employment opportunities, and increasing the quality of life of an individual and a society as a whole [14]. The concept of Industry 4.0 or I4.0 can be summarized starting from mechanization that took place in the beginning during the 1800s' electrification and industrialization during the 1900s, digitization and electronic automation from the 1960s onward, and, lastly, we have reached toward a stage where many technologies seem to be converging and moving toward singularity; the name given to this set of disruptive technologies that have created a great paradigm shift in almost all walks of life is I4.0, comprising the key elements as interoperability and connectivity. This idea was first introduced by the German government toward their industrial strategic initiative and termed these efforts that were toward automation with minimal human intervention as I4.0. It was a strategic initiative undertaken by Germany to transform their massive production processes by automating their industrial manufacturing sector through digitization and thereby paving the way to capitalizing the potential of new ICT technologies availing demand of global markets through individualized and customized products.

I4.0 conceptually is stated as the set of physical systems integrated with ICT components based on the integration of businesses and manufacturing processes inclusive of all participants in the company's value chain which encompasses from suppliers to the end-users who are also known as customers. The technicalities of this new paradigm are addressed by interconnected physical systems via IoT and IIoT related to production systems. The execution system of this whole paradigm is governed by CPS, embedded with electronics systems. These systems are enabled with decentralized control and internet connectivity that enables all the connected devices to this network to gather and exchange real-time data to collect and exchange real-time data, thus enabling the whole interconnected system to be traceable, locate, identify, monitor, and optimize the whole industrial manufacturing process involved in the production systems [3]. These interconnected ICT-enabled large industrial systems have succumbed to several sectors of the economy, ranging from the energy sector in which power generation and distribution are included; apart from them, health care, manufacturing, public sector, transportation, mining, etc., are also influenced by these advancements of technology, and Figure 7.1 illustrates the same.

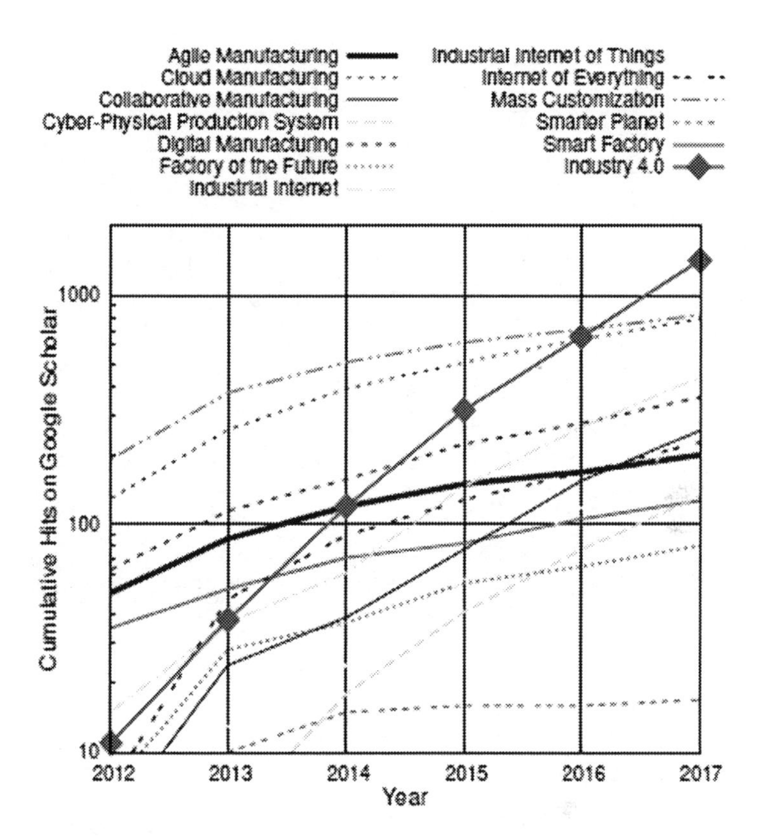

FIGURE 7.1
I4.0 in scientific literature.

Humankind is presently amid hi-tech transformation with the advent of I4.0 since it is essentially changing the way we work and live. It is challenging to foresee in particular how this change is going to influence various businesses and nations but the significance of this new trend in technological advancement can be understood by a number of research articles being published in highly indexed journals as depicted in Figure 7.2. As is depicted in Figure 7.3, I4.0 incorporates several aspects of Industrial Revolution 4.0, which are considered as the building blocks of I4.0 comprising automation, simulation, smart supply chain network, self-organizing, and robot-assisted, augmented work storing its data over cloud, able to make wise decisions with Big Data analytics embedded with machine learning algorithms which provide efficient and effective.

The need for solutions to enterprises with connected intelligence amplify the value of digitization [15]. Machines working with the aid of robots which are built upon Artificial Intelligence are connected together with the help of the internet and the data being created by these machines through interactions

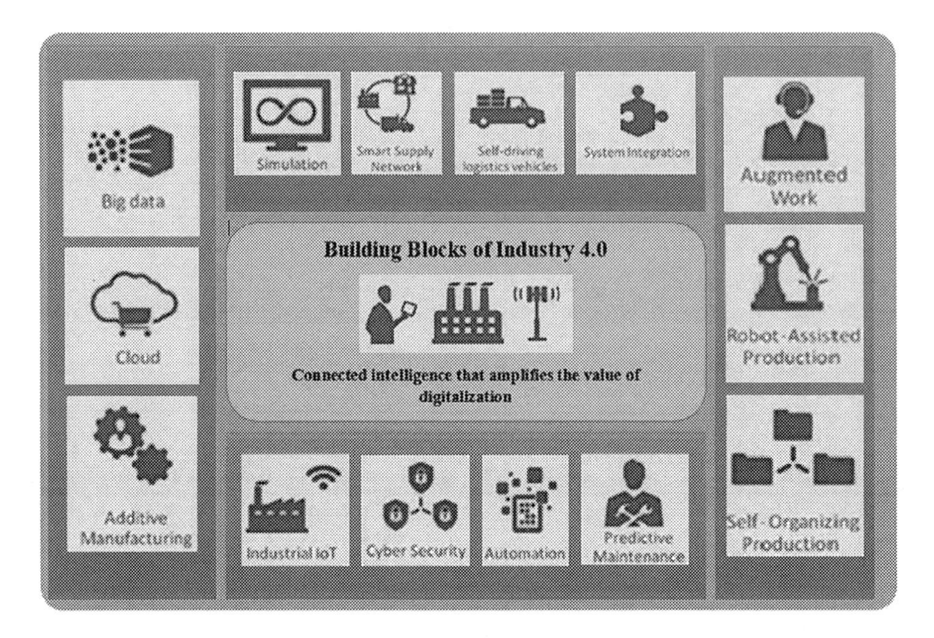

FIGURE 7.2
I4.0 landscape.

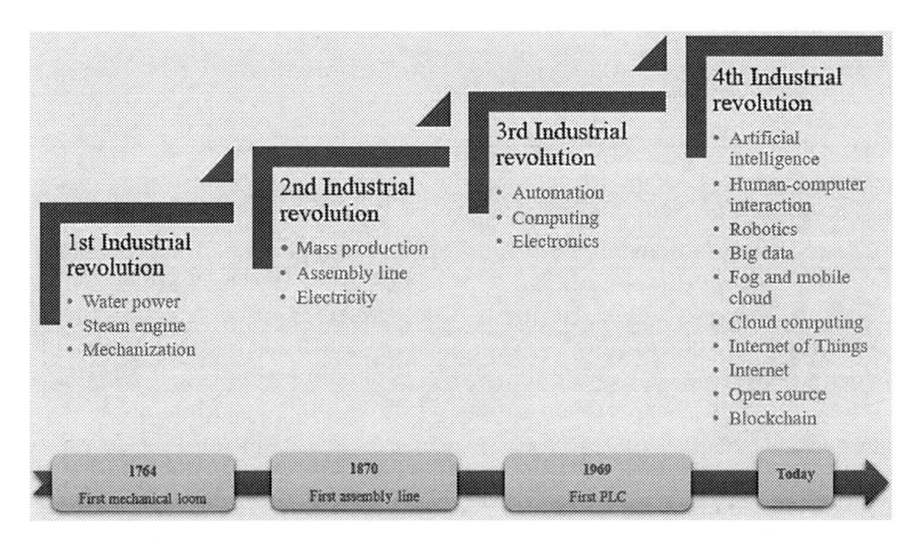

FIGURE 7.3
Industrial revolution enablers.

between them is sent to the cloud, and through the cloud, with the help of machine learning algorithms, the enormous amount data being generated by this plethora of connected devices are made meaningful, thereby providing valuable insights which are helping the organizations to grow, and are indeed getting helpful in making effective and efficient decisions moving organizations toward greater success; as stated, I4.0 aims at "making better things while making things better" connoting new trends in industries which are basically connected with the internet and are termed IIoT. Though they are connected through CPS, there is a risk of data theft and other security concerns which may sabotage the whole network; to overcome this, the network is equipped with state-of-the-art technologies which provide cyber security to all these networks with predictive maintenance. New products are reaching the markets within a shorter period due to this high-tech automation of industries. Customer responsiveness has increased enormously. Mass production has been observed without an actual increase in overall production costs. This paradigm of complex technologies has enabled a flexible and friendlier work environment with efficient use of energy and natural resources. I4.0 has given rise to the concept of smart factories wherein raw materials enter the manufacturing sector and exit as finished products with little or no human intervention.

7.4 I4.0 Enabling Technologies

Key enabling technologies are comprised of physical or digital interface technologies, networking technologies, data processing technologies, and physical process technologies. Physical or digital interface technologies constitute the physical devices layer which includes devices related to IoT, CPS, and visualization technologies. Networking Technologies include cloud computing and blockchain technology. Data processing technologies include simulation and modeling, Machine Learning and AI algorithms, and Big Data Analytics. Physical Process Technologies include 3D printing, advanced robotics, and energy management solutions [14]. As emphasized in Figure 7.3, the key enabling technologies for this whole concept to be a reality are: the internet, cloud computing, Human–Computer Interaction (HCI), Big Data, Artificial intelligence, IoT, Robotics, Fog and Mobile cloud, Augment Reality, cybersecurity, and open-source technologies. These empowering technological advancements are shifting the whole industrial environment by adding a new dimension resulting to bring about a sensational increment in industrial productivity. All these sets of complex technologies can be succinctly summarized as follows.

Internet: The Internet is defined as a globally interconnected network of physical devices, also known as nodes, for the purpose of communication based on rules that govern the communication over computer networks with the help of TCP/IP protocol. Devices spread over the globe are connected using transmission control over internet protocol. The internet is a pool of unlimited resources pertaining to information and services, for instance, hypertext documents that are linked together; apart from that, it also provides support for electronic mails, internet telephony, and file sharing over the World Wide Web. The devices connected globally are tied together with numerous networking technologies, Thus without the internet, we can't imagine I4.0 happening a reality as it provides the basic infrastructure and also is the basis for this whole concept of I4.0, its acts as a backbone for this whole phenomenon [16,17].

IoT: Apart from the internet, the other key technology enablers are IoT; according to one of the estimations, by the end of 2020, it is expected that around 50 billion physical devices will be interconnected with the ability to communicate with each other with minimal human intervention. This technological enhancement can drastically improve the quality of life by enabling smart real-life applications ranging from home, city, and healthcare to CPS at large. IoT is envisioned as a three-layer architecture wherein each layer is associated with specific responsibilities and functionalities, namely the perception layer, the networking layer, and the application layer, respectively. The perception layer constitutes physical objects that can sense the environment's data which can be textual, audio, video, and images as well. Thus, the sensed data are transmitted to the next layer comprising networking and communication technologies, and from this layer, the perceived data are pushed to the next level of the layered architecture, namely the application layer. Here, the data gathered are made available to individuals and organizations from where they can take decisions and proceed further [18,19]. Hence, IoT is basically an extension of the communication between man to man to objects communicating with minimal or no human intervention with the assistance of robots.

Robotics: The word "robot" is derived from the Slavic word "robota" meaning slave or servant. Robots are used in the manufacturing sector to assemble, pack, and transport. Robotics is an emerging research area comprising computer science and several engineering disciplines. Robots are defined as machines that can sense think and act in a real world with the help of algorithms designed and enabled with Artificial Intelligence, thus paving the way for the rapid production of products with the help of additive manufacturing technologies [20–23]. Additive manufacturing technologies provide the basis for 3D printing.

3D Printing: The process of using machines to create 3D objects is known as 3D printing. It comprises processes where materials are deposited, joined, and solidified under computer control with the help of computer-aided design (CAD). CAD helps designers with increased productivity by availing them in creating, modifying, analyzing, and optimizing the design of physical objects. It paves the way for creating custom parts which can help in large manufacturing areas with rapid replacement of physical parts of a machine involved in manufacturing processes [24]. As the enormous amount of data will be generated through the communications between physical devices, there is a need to make this data fruitful for the betterment of the businesses; this is done with the help of Big Data Analytics.

Big Data Analytics: The term Big Data is characterized by 3 Vs with a supplement of 2 more Vs, the first 3 Vs connote volume, velocity, and variety, and the latter 2 Vs connote veracity and value. The mechanism of processing large amounts of data to gain meaningful insights which further can help organizations with better decision-making techniques is termed Big Data Analytics. In I4.0, Big Data Analytics plays a vital role as it provides the mechanisms to capture, store, transfer, curate, search, analyze, and visualize with the intent to provide privacy and security to the data owned by an organization. In Big Data, data are captured via system logs, cameras, RFIDs, sensors, GPS, ERP, and social media. The captured data are further processed using Data Analytics by applying descriptive, predictive, and prescriptive methods. In the whole context of analyzing the data, the data captured through various sources of an organization is stored and analyzed – the descriptive-analytical method describes the past data about a product, the predictive-analytical method provides the insight into the future of the product, and, last but not least, the prescriptive-analytical method provides insight to prepare the product for the future; in this whole analytics process, if the data that are captured are incorrect, then the whole process will be futile, and hence, extra care is taken by the organizations to the data accuracy which is known as veracity that in turn creates value to the product which is also considered as attributes of Big Data [25,26]. The data transmission and storage need also is enormous in this complex network of devices hence, cloud computing provides services for this task.

Cloud Computing: This term basically connotes data centers wherein the data are stored but in broader perspective cloud computing provides services in a pay-as-you-go mode, wherein the user or an organization needs to pay only for the services they acquire from the cloud service providers. Cloud computing provides in terms of hardware, platform, infrastructure, and software as a service. All the services pertaining to hardware are termed "Hardware as a Service" (HaaS), Platform as a Service (PaaS), Infrastructure as a Service (IaaS), and Software as a service (SaaS), respectively. Cloud computing facilitates the lending of computing resources comprising storage, computational power, hardware, and software as a service in real time with reduced interaction with the service provider, thereby availing scalability, responsiveness, and considerable economic gain. Advantages of cloud computing include services pertaining to response time, connectivity, bandwidth, cost, and connection availability [27–29]. Thus, to ease the latency and response time in communication, mobile and fog computing technologies are introduced.

Fog Computing: Cloud computing services transferred to the nearest cloud which is known as an edge network is termed fog computing. Fog computing facilitates to balance the load between local and end devices and data centers with low latency rates, fast response with increased performance, and scalability of the whole system. Thus, fog computing plays a vital role in minimizing the response time between the user and the data center. Apart from the aforesaid technologies, Augmented reality, Simulation, open-source software technologies, HCI, and AI also add a significance to this whole network of I4.0 to analyze the physical systems and to inform the concerned authorities with real-time data thereby reducing the costs and improving the efficiency of the overall system and this whole mechanism entails better performance and better real-time decision-making strategies with which organization can flourish with better economic gains. As the whole network is connected through CPS systems, the security of the whole system will be the primary concern; hence, to avail this, cyber security plays a vital role in protecting and safeguarding the whole system from security breaches, and blockchain technologies provide best solutions to address these security-related issues [30–33].

7.5 Industry 4.0 Issues and Challenges

In the past, manufacturing systems were closed and their security was entrusted as they were placed in isolation and their physical access was controlled by specific individuals assigned to look and operate over them but now, conversely with the advent of I4.0 technologies, the manufacturing systems have been replaced with technological advancements wherein modern manufacturing sector has been automated with machines that are interconnected with embedded smart devices such as sensors and actuators which are connected either by wired or wireless network systems to process data using specific protocols which are inadequate to provide protection against cyber threats. These inadequacies have made the manufacturing systems vulnerable to cyberattacks claiming unauthorized users to access the data pertaining to organizations [34]. Critical assets in I4.0 are industrial control systems, IIoT, IIoT gateways, sensors, and actuators [35]. IIoT gateways manage many communication technologies involving both wired and wireless technologies with the application of various protocols. The threats engrossed in this whole scenario include loss of data confidentiality, data integrity, and availability [36]. As evident from the scientific literature, there is a substantial amount of investment in various industrialized sectors such as defense, security, aerospace, chemical engineering, construction, automotive, electronics transportation, industrial manufacturing, etc. This has provided a great impetus for industrial growth wherein new products are reaching markets within a shorter period of time; customer responsiveness has increased enormously and massive production has become possible without an increase in overall production costs. This has enabled a flexible and friendlier work environment with efficient use of energy and natural resources. Though I4.0 has paved the way for the automation of the industrialized manufacturing sector by automating the production units in factories and making these factories smart. Smart factories are the places where the raw materials are entered into the production units and are processed into finished products and leave from the production unit till it reaches the end-user or customer with little or no human intervention. This whole automation of the value chain is made possible with a varied level of technologies involved in this whole complex process. As the whole process is automated with the CPS, there is a great risk of security which if neglected may lead to failure of the system, thereby making the whole process futile. Enabling and providing a tight security mechanism is considered to be one of the greatest challenges in this whole paradigm of I4.0 as depicted in Figure 7.1. Apart from these security issues, the other challenges include interoperability, because of the involvement of heterogeneous networks.

The automation of industrial operations ranging from designing and manufacturing to supply chain and service maintenance with the help of interconnected and intelligent CPS is envisioned as I4.0. In this whole plethora of connected world, exponential growth has been seen in cyberattacks affecting the performance of businesses negatively. A study conducted by Engineering Employers' Federation has revealed that manufacturers using CPS based on I4.0 reported that about 48% of them have been the soft target of cybersecurity attacks, and among these, 48% of them have claimed to have severe financial losses [36].

Since the manufacturing sector involves machines connected to the network via smart devices and computers, they are easy targets of cyberattacks which can harm the organizations in terms of sabotaging critical infrastructure which comprises the whole machinery connected to this network. Apart from it, there can be Denial of Service (DoS) attacks on both networks and computers involved in this system; lastly, it cannot be ignored that the data theft pertaining to trade secrets and Intellectual Property Rights is also a great concern for the organizations; all these factors have a considerable impact on organizations' growth in terms of economical means.

The impact of I4.0 has broadened every aspect of human life ranging from individual, societal, entrepreneurial, governmental, and educational sectors as well; hence, the foremost challenges and issues faced by this new paradigm are regulations as without proper regulatory authorities, the system can't run so can't I4.0; apart from regular authorities, there is a need for standardization as well, resulting in the development of new standards pertaining to communication, industries, products, services, business models pertaining to markets. To make industries ready for I4.0, many initiatives have been taken by several countries but still, there are issues pertaining to their readiness in terms of cost, sustainability, energy efficiency, economy, work environment, and skill development of the employees working in an organization [13,37,38] (Figure 7.4).

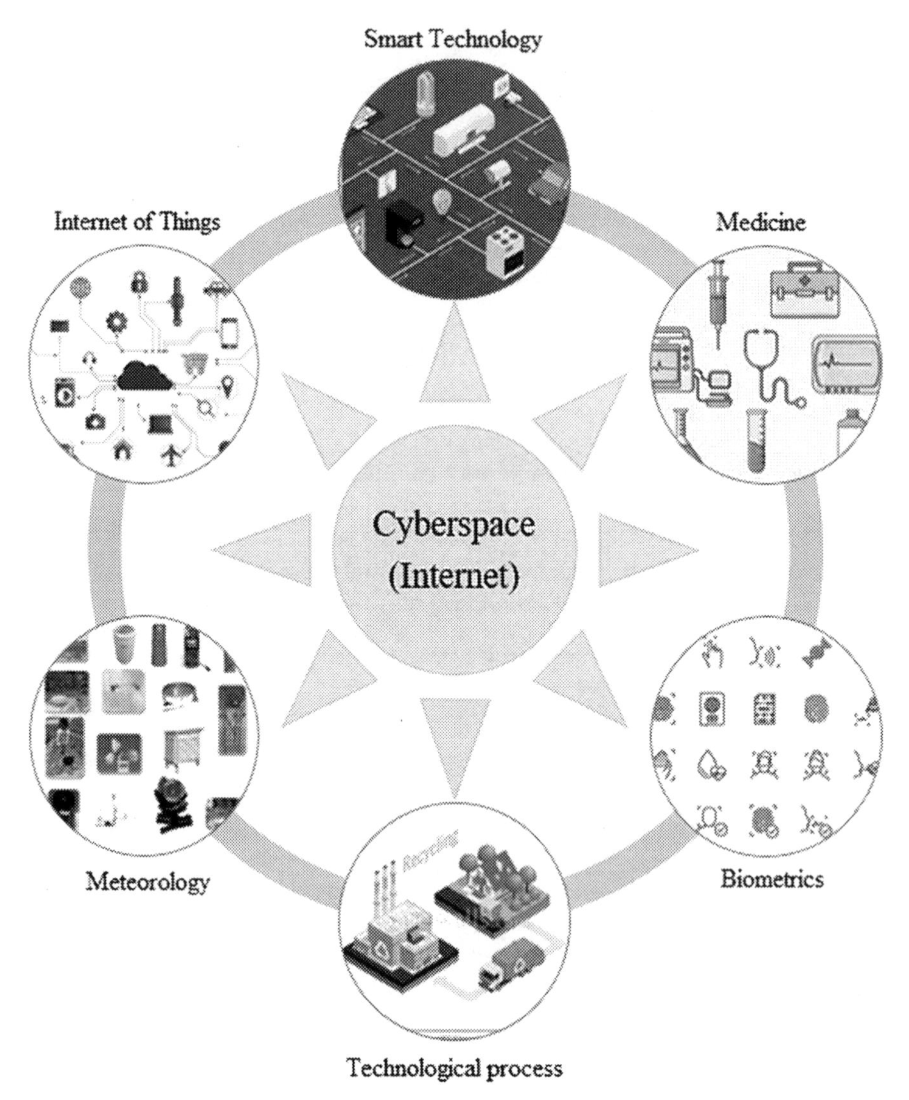

FIGURE 7.4
Cyber-physical systems landscape.

7.6 Discussion

The world is as of now in the midst of a technological change composed of various paradigms and is thus compelling industries toward a paradigm shift that will generally change the manner in which we live and work. It is hard to anticipate absolutely how this change will affect various business undertakings and populations. Nevertheless, the realization is that this change is not usual for anything that has been viewed and experienced before because of the sheer pace and expansiveness of these changes. Many consider these to be a guiding phenomenon for a new period of development, change, and opportunity. The repercussion it brings for assembling firms is compelling countries and governments to rethink how the assembling business works and adds to monetary development. Countries are presently expanding their attention on creating propelled producing capacities by putting resources into cutting-edge frameworks and quality instructions. These are finished by encouraging the change of the assembling businesses and their organizations by employing innovative technologies resulting in their financial prosperity. The main mechanical upheaval began with the appearance of steam and water control empowering the motorization of creation firms, which resulted in the first industrial revolution (IR 1.0); at this stage, only the textile industry was the prime actor of influence shifting manual handloom industries toward mechanization. The second mechanical unrest was driven by electric power and mass assembling strategies, connoting the second industrial revolution (IR 2.0); this triggered the discovery of electricity enabling electrification and industrialization, moving the whole manufacturing process toward massive production enabled with assembly lines. The invention of computers embedded with electronics resulted in great change in industries resulting in the third industrial revolution acronymized as IR 3.0; this paved the way toward digitization and electronic automation with the advent of microcontroller technology which created a way to design programmable machines. The fourth industrial revolution with the invention of the internet makes communication possible not only between human beings but also between objects that are connected through this seamless network of things generating a huge amount of data, enabling technologies to provide better and more efficient solutions with the help of Big Data analytics. The fourth industrial revolution or Industry 4.0 changed the whole paradigm of how production systems of goods and services were planned, manufactured, utilized, and worked just as how they are kept up and adjusted. It is likewise changing the activities, forms, storing networks, executives, and vitality impression of production lines. Industry 4.0 changes the worldwide scene of assembling rivalry, decreasing the relative upper hand of ease that depends on modest work. Countries and assembling firms that lead in grasping Industry 4.0 innovations and procedures have increased worldwide competitors.

Incorporation of ICT-enabled services is revolutionizing whole massive production based on demands by the global markets and whole value chain by decentralizing and providing better monitoring facilities tracing the whole industrial production landscape. I4.0 and technological base in CPS with strong internet connectivity forced the change from whole traditional classical hierarchical automation to self-organized cyber-physical production systems [26].

7.7 Conclusion

The development that took place in industrial manufacturing sector can be understood by tracing back history through the 1800s wherein the discovery steam having the power to move things had given rise to mechanical power and enabled mechanization, thereby revolutionizing the whole manual work of manufacturing through mechanization in this era and mostly the textile industry was influenced; later on, the IR 2.0 was triggered with the discovery of electricity, enabling electrification and industrialization that moved the whole manufacturing process toward massive production; this process continued until the invention of computers embedded with electronic chips, which paved the way for the invention of electronic computers enabling digitization and electronic automation; to this continuum, the invention of the internet gave an impetus to whole human lives toward global connectivity through computers and thus communication among humans continued with a paradigm shift when the communication concept was imposed on machines, that is, objects communicating with each other through sensors combined with infrastructure provided by CPS, creating enormous amount of data, and exchanging this data through the network with processing power and ICT-enabled services revolutionizing whole industrial manufacturing process with the introduction of IIoT with minimal human intervention. In this chapter, the introduction, concept, significance, challenges, and issues of I4.0 are presented. Though the whole phenomenon is very comprehensive and complex, it can be summarized as follows.

IoT applied to the industrial or manufacturing sector and will pave the way to a completely new generation of manufacturing systems based on CPS enabled with ICTs. This chapter has elucidated the concept of Industry 4.0 embodied by connectivity and digitized production. Thus, the conclusion is that more deliberation is given to production systems, management involved in it, and economies participating for improved productivity. The transformation which is evident from the recent technological advancements and their contributions to revolutionizing the productivity of industries is eliminating the boundaries between the virtual and physical worlds, assimilating workers, smart machines, smart products, production systems, and processes into its paradigm.

7.8 Future Work

I4.0 is envisioned to be a set of complex technologies with a great potential that can be instrumental in providing good quality of life. Though it is embedded with great potential to serve and improve the quality of life, there are many challenges that need to be addressed in terms of cyber security, feasibility, scalability, and interoperability as there are many different systems and various industrial vendors involved in this whole exercise.

References

1. P. Prisecaru, "Challenges of the fourth industrial revolution," *Knowledge Horizons. Economics*, vol. 8, no. 1, p. 57, 2016.
2. A. Pereira and F. Romero, "A review of the meanings and the implications of the Industry 4.0 concept," *Procedia Manufacturing*, vol. 13, pp. 1206–1214, 2017.
3. G. Aceto, V. Persico, and A. Pescapé, "A survey on information and communication technologies for Industry 4.0: State of the art, taxonomies, perspectives, and challenges," *IEEE Communications Surveys & Tutorials*, 2019.
4. C. Santos, A. Mehrsai, A. Barros, M. Araújo, and E. Ares, "Towards Industry 4.0: An overview of European strategic roadmaps," *Procedia Manufacturing*, vol. 13, pp. 972–979, 2017.
5. K. Zhou, T. Liu, and L. Zhou, "Industry 4.0: Towards future industrial opportunities and challenges," in *2015 12th International Conference on Fuzzy Systems and Knowledge Discovery (FSKD)*, 2015: IEEE, pp. 2147–2152.
6. G. Reischauer, "Industry 4.0 as policy-driven discourse to institutionalize innovation systems in manufacturing," *Technological Forecasting and Social Change*, vol. 132, pp. 26–33, 2018.
7. J. Wu, S. Guo, H. Huang, W. Liu, and Y. Xiang, "Information and communications technologies for sustainable development goals: State-of-the-art, needs and perspectives," *IEEE Communications Surveys & Tutorials*, vol. 20, no. 3, pp. 2389–2406, 2018.
8. R. Davies, "Industry 4.0 digitalization for productivity and growth. European parliamentary research service," *Members' Research Service PE*, vol. 568, 2015.
9. J. I. R. Molano, J. M. C. Lovelle, C. E. Montenegro, J. J. R. Granados, and R. G. Crespo, "Metamodel for integration of internet of things, social networks, the cloud and industry 4.0," *Journal of Ambient Intelligence and Humanized Computing*, vol. 9, no. 3, pp. 709–723, 2018.

10. F. Shrouf, J. Ordieres, and G. Miragliotta, "Smart factories in Industry 4.0: A review of the concept and of energy management approached in production based on the Internet of Things paradigm," in *2014 IEEE International Conference on Industrial Engineering and Engineering Management*, 2014: IEEE, pp. 697–701.
11. S. Weyer, M. Schmitt, M. Ohmer, and D. Gorecky, "Towards Industry 4.0-standardization as the crucial challenge for highly modular, multi-vendor production systems," *IFAC-Papersonline*, vol. 48, no. 3, pp. 579–584, 2015.
12. O. Elijah, T. A. Rahman, I. Orikumhi, C. Y. Leow, and M. N. Hindia, "An overview of Internet of Things (IoT) and data analytics in agriculture: Benefits and challenges," *IEEE Internet of Things Journal*, vol. 5, no. 5, pp. 3758–3773, 2018.
13. B. A. Kadir, O. Broberg, and C. S. da Conceição, "Current research and future perspectives on human factors and ergonomics in Industry 4.0," *Computers & Industrial Engineering*, p. 106004, 2019.
14. G. Culot, G. Nassimbeni, G. Orzes, and M. Sartor, "Behind the definition of Industry 4.0: Analysis and open questions," *International Journal of Production Economics*, vol. 226, 2020, doi: 10.1016/j.ijpe.2020.107617.
15. E. Manavalan and K. Jayakrishna, "A review of Internet of Things (IoT) embedded sustainable supply chain for industry 4.0 requirements," *Computers & Industrial Engineering*, vol. 127, pp. 925–953, 2019.
16. M. Nikkhah, R. Guérin, and M. Nikkhah, "Migrating the internet to IPv6: An exploration of the when and why," *IEEE/ACM Transactions on Networking (TON)*, vol. 24, no. 4, pp. 2291–2304, 2016.
17. S. Neumayer, G. Zussman, R. Cohen, and E. Modiano, "Assessing the vulnerability of the fiber infrastructure to disasters," *IEEE/ACM Transactions on Networking*, vol. 19, no. 6, pp. 1610–1623, 2011.
18. M. A. Al-Garadi, A. Mohamed, A. Al-Ali, X. Du, I. Ali, and M. Guizani, "A survey of machine and deep learning methods for Internet of Things (IoT) security," *IEEE Communications Surveys & Tutorials*, pp. 1–1, 2020, doi: 10.1109/comst.2020.2988293.
19. A. Nauman, Y. A. Qadri, M. Amjad, Y. B. Zikria, M. K. Afzal, and S. W. Kim, "Multimedia internet of things: A comprehensive survey," *IEEE Access*, vol. 8, pp. 8202–8250, 2020, doi: 10.1109/access.2020.2964280.
20. D. Bhargava, "Intelligent Agents and Autonomous Robots," in *Rapid Automation: Concepts, Methodologies, Tools, and Applications*: IGI Global, 2019, pp. 1134–1143.
21. R. J. Alattas, S. Patel, and T. M. Sobh, "Evolutionary modular robotics: Survey and analysis," *Journal of Intelligent & Robotic Systems*, vol. 95, no. 3–4, pp. 815–828, 2019.
22. T. Tangiuchi et al., "Survey on frontiers of language and robotics," *Advanced Robotics*, vol. 33, no. 15–16, pp. 700–730, 2019, doi: 10.1080/01691864.2019.1632223.
23. L. Junjun, L. Zhijun, F. Chen, A. Bicchi, Y. Sun, and T. Fukuda, "Combined sensing, cognition, learning and control to developing future neuro-robotics systems: A survey," 2019.
24. P. D. Chatzoglou and V. N. Michailidou, "A survey on the 3D printing technology readiness to use," *International Journal of Production Research*, vol. 57, no. 8, pp. 2585–2599, 2019, doi: 10.1080/00207543.2019.1572934.

25. S. R. Salkuti, "A survey of big data and machine learning," *International Journal of Electrical and Computer Engineering (IJECE)*, vol. 10, no. 1, pp. 575–580, 2020, doi: 10.11591/ijece.v10i1.
26. L. D. Xu and L. Duan, "Big data for cyber physical systems in industry 4.0: A survey," *Enterprise Information Systems*, vol. 13, no. 2, pp. 148–169, 2019, doi: 10.1080/17517575.2018.1442934.
27. G. Aceto, V. Persico, and A. Pescapé, "Industry 4.0 and health: Internet of things, big data, and cloud computing for healthcare 4.0," *Journal of Industrial Information Integration*, vol. 18, p. 100129, 2020, doi: 10.1016/j.jii.2020.100129.
28. M. Ramirez-Peña, A. J. Sánchez Sotano, V. Pérez-Fernandez, F. J. Abad, and M. Batista, "Achieving a sustainable shipbuilding supply chain under I4.0 perspective," *Journal of Cleaner Production*, vol. 244, p. 118789, 2020, doi: 10.1016/j.jclepro.2019.118789.
29. R. S. Nakayama, M. De Mesquita Spínola, and J. R. Silva, "Towards I4.0: A comprehensive analysis of evolution from I3.0," *Computers & Industrial Engineering*, vol. 144, p. 106453, 2020, doi: 10.1016/j.cie.2020.106453.
30. Z. A. Almusaylim and N. Zaman, "A review on smart home present state and challenges: Linked to context-awareness internet of things (IoT)," *Wireless Networks*, 2018, doi: 10.1007/s11276-018-1712-5.
31. S. Kok, A. Abdullah, N. JhanJhi, and M. Supramaniam, "Prevention of crypto-ransomware using a pre-encryption detection algorithm," *Computers*, vol. 8, no. 4, p. 79, 2019.
32. Z. A. Almusaylim and N. Jhanjhi, "Comprehensive review: Privacy protection of user in location-aware services of mobile cloud computing," *Wireless Personal Communications*, pp. 1–24, 2019.
33. M. Lim, A. Abdullah, and N. Jhanjhi, "Performance optimization of criminal network hidden link prediction model with deep reinforcement learning," *Journal of King Saud University-Computer and Information Sciences*, pp. 1202–1210, 2019.
34. D. Wu et al., "Cybersecurity for digital manufacturing," *Journal of Manufacturing Systems*, vol. 48, pp. 3–12, 2018.
35. D. Sullivan, E. Luiijf, and E. J. Colbert, "Components of Industrial Control Systems," in *Cyber-Security of SCADA and other Industrial Control Systems*: Springer, 2016, pp. 15–28.
36. A. Corallo, M. Lazoi, and M. Lezzi, "Cybersecurity in the context of industry 4.0: A structured classification of critical assets and business impacts," *Computers in Industry*, vol. 114, 2020, doi: 10.1016/j.compind.2019.103165.
37. M. N. Sadiku, Y. Wang, S. Cui, and S. M. Musa, "Industrial internet of things," *International Journal of Advances in Scientific Research and Engineering*, vol. 3, pp. 486–489, 2017.
38. R. Alguliyev, Y. Imamverdiyev, and L. Sukhostat, "Cyber-physical systems and their security issues," *Computers in Industry*, vol. 100, pp. 212–223, 2018.

8

IIoT-Based Secure Smart Manufacturing Systems in SMEs

S. Karthi, N. Narmatha and M. Kalaiyarasi
V.S.B. Engineering College, Karur, India

CONTENTS

DOI: 10.1201/9781003203087-8

8.1 Introduction

The Internet of Things (IoT) is a new innovation that utilizes the Internet to connect actual gadgets or "things." Home apparatuses and mechanical hardware are examples of actual gadgets. These contraptions can give imperative information and empower buyers to get an assortment of administrations by using legitimate sensors and correspondence organizations. Controlling the energy consumption of buildings in a smart way, for example, allows for lower energy expenses. The IoT may be used in a variety of industries, including manufacturing, logistics, and construction. In addition to environmental monitoring, the IoT is increasingly used in healthcare systems and services, drone-based services, and smart energy management in buildings. The determination of IoT parts like detector gadgets, correspondence conventions, information stockpiling, and registering should be fitting for the planned application while making an IoT application, which is the initial stage in designing IoT systems. These all application in Industry is technically termed as Industrial Internet of Things (IIoT). The IIoT or the Fourth Industrial Revolution or Industry 4.0 is, for the most part, the name given to the use of IoT development in a business setting. Industry 4.0 is a wide view with clear constructions and remarkable plans, fundamentally described by the crossing over of real current benefits and automated developments in supposed advanced real structures. Small- to medium-sized enterprises (SMEs) are endeavors that are ordered based on the size of speculation and they exist in both the Manufacturing and Services Sectors of the economy. The IIoT has the greatest impact on smart manufacturing systems (SMSs) in SMEs.

8.2 Internet of Things

The IoT alludes to an approach of interdependent, web-related things that can gather and move information over a faraway relationship without human intercession [1]. The IoT represents the confederation of unfeigned articles, so known as, "things" – that are introduced with actuators, programming, and various advancements that are utilized to partner and exchange data with various contraptions and structures over the Internet. The IoT, also called the Internet of Everything or the Industrial Internet, is another advancement perspective visualized as an overall association of appliances and contraptions outfitted to partner with each other [2]. The IoT is portrayed as the billions of real contraptions all throughout the planet that are as of now connected with the Internet, all social events, and exchanging information.

Interfacing up this heap of different things and adding actuators to them adds a degree of mechanized information to gadgets that would be by and large numbskull, engaging them to pass on constant information without including an individual. Figure 8.1 shows the features of the IoT. The IoT is making the outside of our in-general ecological variables more shrewd and more accessible, consolidating the significant level and real universe [3]. The upsides of the IoT for employment rely upon the specific execution; ability and adequacy are commonly top thoughts. The thinking is that attempts should move toward much information about their things and their inner frameworks, and a more vital capacity to make alterations likewise. The IoT pledges to build our present conditions – our accommodation and working environments and automobiles – savvier, more quantifiable, and more effusive. Autonomous vehicles and smart metropolitan regions could change how we assemble and deal with our courtyard. Security is perhaps the best issue with the IoT. These sensors are assembling, a large part of the time, incredibly fragile data. An enormous number of IoT gadgets give tiny plans to the fundamentals of wellbeing, like encoding information becoming extremely still. It expects that gadgets ought to have remarkable passwords, that associations will give a common asset so anybody can delineate a shortcoming (and that these will be circled back to), and that creators will expressly tell long-lasting gadgets will get safety revives. It's an unassuming summary, yet a beginning. Right when the price of manufacturing splendid things is set off immaterial, these issues will simply end up being more expansive and obstinate. The IoT is making and keeps being the most recent, most advanced idea in the IT world. Over the previous ten years, the phrase

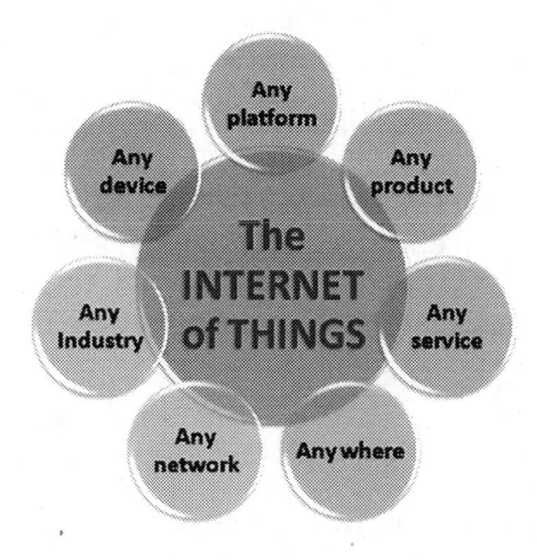

FIGURE 8.1
Features of Internet of Things.

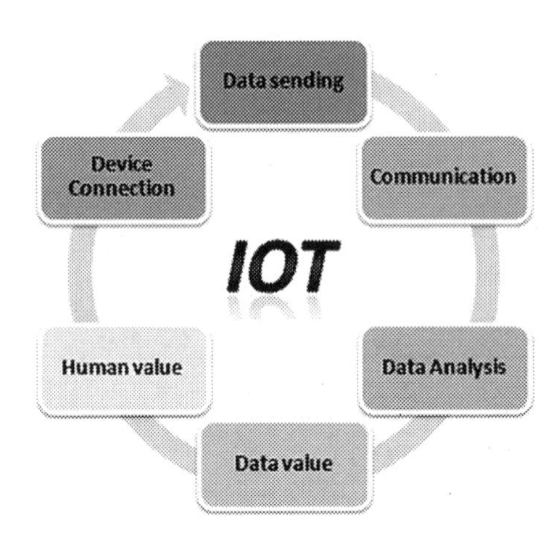

FIGURE 8.2
IoT.

"Internet of Things" has stood separated by widening the view of a general foundation of coordinated certified articles, empowering whenever, any spot is available for anything and not just for anyone [4]. Figure 8.2 shows the IoT.

8.3 Industrial Internet of Things

The IIoT or the Fourth Industrial Revolution or Industry 4.0 is, for the most part, a label given to the usage of IoT advancement in a professional setting. The idea is similar to the purchaser of IoT gadgets in the convenience; anyway, for the current situation, the fact is to use a mix of actuators, far-off associations, large information, AI, and assessment to check and progress industrial cycles [3]. The IIoT is defined as the interconnected actuators, instruments, and different gadgets arranged along with PCs' modern applications, including assembling and energy the executives. This network takes into account information assortment, trade, and investigation, conceivably working with upgrades in usefulness and productivity just as other financial advantages. The IIoT is an advancement of a Distributed Control System (DCS) that considers a more serious level of computerization by utilizing distributed processing to refine and upgrade the interaction controls. The IIoT is empowered by innovations like network safety, distributed computing, edge registering, versatile advances, machine-to-machine (M2M), 3D printing, progressed modern mechanics, huge information, the IoT, RFID innovation,

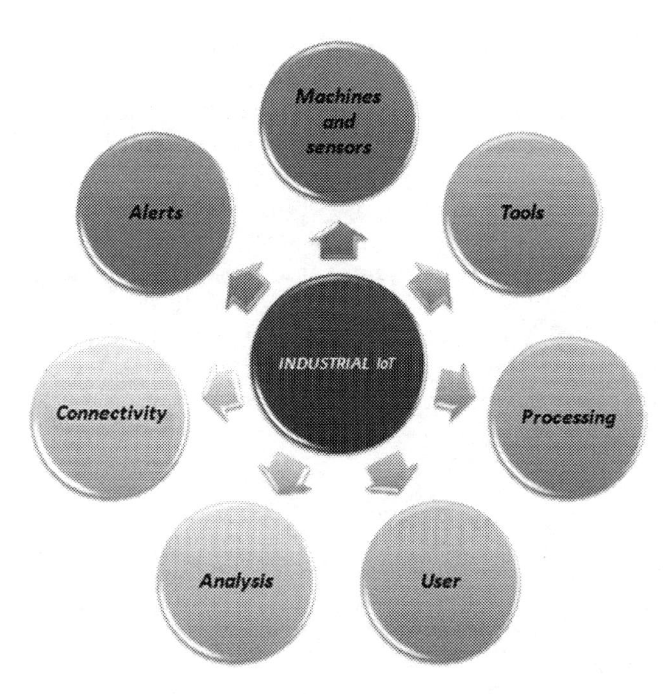

FIGURE 8.3
Industrial IoT.

and psychological processing [5]. The IIoT addresses the increase and benefit of the IoT in mechanical regions and supplications. With a rigid core interest in M2M resemblance, huge data, and AI, the IIoT engages in adventures and tries to have superior capability and constancy in their undertakings. The IIoT wraps mechanical supplications, together with progressed mechanics, clinical gadgets, and programming portrayed creation measures [6]. Figure 8.3 shows the IIoT. The IIoT is becoming rapidly because of expanding arrangement and mix of keen actuators, apparatus, gadgets, and programming utilizing wired or remote organizations. Through this coordinated equipment programming approach, modern practices will improve essentially, bringing about mechanical insight for more productive assembling [7]. The IIoT goes past the ordinary shopper machines and interconnecting of actual gadgets, for the most part, connected with the IoT. What makes it undeniable is the junction point between information technology (IT) and operational technology (OT). OT alludes to the systems administration of functional cycles and industrial control systems (ICSs), including human–machine interfaces (HMIs), supervisory control and data acquisition (SCADA) frameworks, Distributed Control System (DCSs), and programmable logic controllers (PLCs). Concerning the Fourth Industrial Revolution, the IIoT is fundamental to how advanced real structures and creation measures are set to swap by dint of huge data and assessment. Persistent data from actuators and different information provenance assist current devices and establishments in their "dynamic" in preparing pieces of information and unequivocal exercises. It is hence moreover authorized to engage and automate tasks that earlier automatic insurrections couldn't oversee. In more broad environs, the IIoT is critical to making use of instances recognized with related natural frameworks or conditions, for instance, how non-rural bodies set off smart urban areas and production lines become smart manufacturing plants. The IIoT likewise further develops office the executives. Assembling gear is defenseless to wear and tear, which can be exacerbated by specific conditions in a processing plant. Sensors can screen vibrations, temperature, and different components that may prompt imperfect working conditions [8].

8.4 IIoT Architecture

IIoT frameworks are normally considered as layered particular engineering of computerized innovation [5].

- Content layer
- Service layer
- Network layer
- Device layer

Figure 8.4 shows the layers of IoT architecture.

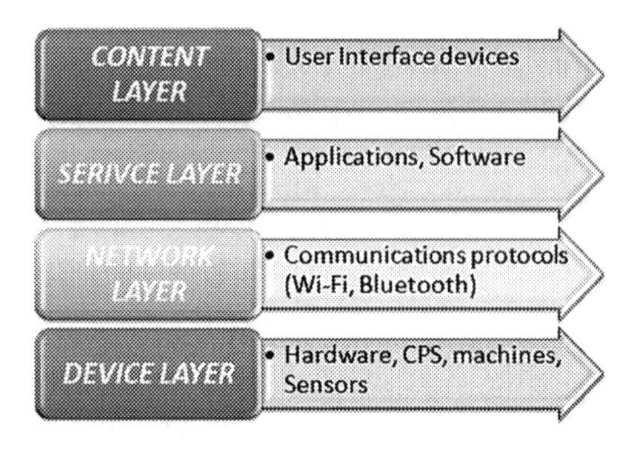

FIGURE 8.4
Layers of IoT architecture.

8.4.1 Device Layer

The device layer alludes to the actual parts: CPU actuators or devices.

8.4.2 Network Layer

The network layer comprises actual organization transports, distributed computing, and correspondence conventions that total and transmit the information to the service layer.

8.4.3 Service Layer

The service layer comprises uses that control and consolidate data into information that can be shown on the driver dashboard.

8.4.4 Content Layer

The highest layer of the heap is the content layer and it is also called the User Interface.

8.5 Benefits of the IIoT

The IIoT is the use of associated brilliant gadgets to screen, computerize, and anticipate a wide range of mechanical cycles and results. These innovations offer everything from improved specialist securities through plant floor checking frameworks to the prescient support prospects as of now reforming the armada the executives business. The far-reaching execution of such frameworks helps the path producers, logistics networks, and stockroom chiefs work all the more successfully [9]. With the IIoT, information-driven bits of knowledge power more prominent outcomes.

For some organizations, this can mean:

- More prominent energy proficiency
- Decreased expenses
- Better quality items
- Further developed dynamic potential
- Less hardware vacation

These advantages are reasons why the IIoT, as a whole, is so appealing. As of now, many organizations across ventures are now utilizing it to have an incredible impact.

8.6 Challenges of Widespread IIoT Implementation

As of now, comprehensive understanding is part of the more extensive difficulties in dealing with a whole modern IoT system with an array of helpful gadgets. Similarly, with any organized gadget, IIoT parts are to be available for network protection chances. In the meantime, the utilization of these gadgets to fill their latent capacity needs pre-arranging and appraisal [9].

While carrying out your IIoT framework, think about these normal difficulties in widespread IoT achievement:

- Inability to adjust KPIs to justifiable calling objections
- Ill-advised authoritative arrangement
- Absence of IoT occurrence
- IoT privacy dangers

The risks of not viewing these difficulties critically can have something other than money-related dangers. A self-governing machine can think twice about well-being like an automobile that has been hacked or contaminated with malware. Beating a large number of these difficulties requires running a network protection hazard evaluation at steady spans in the course of the IIoT gadget's wheel of life and preparing staff throughout the appropriate execution.

8.7 Industry 4.0

The underlying three industrial revolutions are portrayed as being handled by programmed creation expecting water and steam power, usage of mass work and electrical power, and the utilization of voltaic, robotized creation independently. While the intended Industry 4.0 was first put forward in 2011 concerning the target of cultivating the German economy, this industrial rebellion is depicted by its dependence on the usage of CPS furnished for correspondence accordingly and of manufacturing self-administering, not really settled to increase mechanical capability, productivity, security, and straightforwardness [10]. Industry 4.0 is the progressed change of gathering and allied endeavors and avail formation actions. Industry 4.0 is utilized similar to the fourth mechanical distress and it addresses more levels in present-day association and force. Industry 4.0 is a fantasy and thought operating with credential constructions, standardization, and even interpretations on the move. Most Industry 4.0 drives are starting stage forecasts with a confined degree. Most digitization and digitalization attempts, in fact, happen with respect to third and shockingly second current bombshell progresses/goals. In recognizing Industry 4.0, it is fundamental

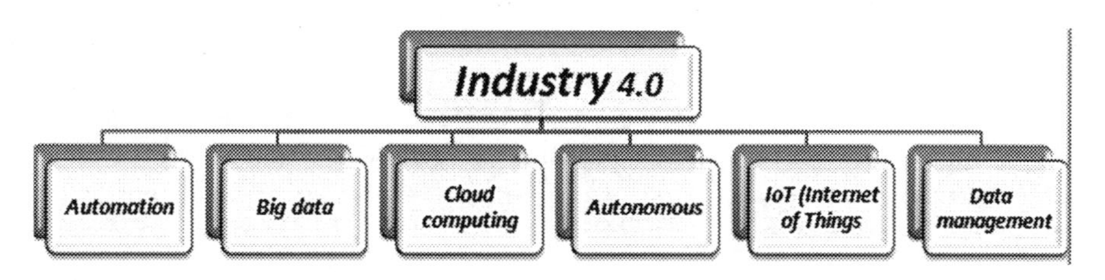

FIGURE 8.5
Industry 4.0.

to see the full worth fetters which consolidate provisioners and the beginning stages of the particulars and parts required for numerous sorts of sharp collecting, the beginning-to-end progressed stock organization, and the last motive of all gatherings, that is, rewarding slight minds to the extent of rewarding center individual advances and participants: the end client. Industry 4.0 is the information heightened change of gathering (and connected endeavors) in a related surrounding of huge information, people, measures, organizations, structures, and IoT-enabled current benefits with the age, impact, and manipulation of imperative data and information as a path and plan to recognize smart industries and natural frameworks of mechanical turn of events and composed exertion. Consequently, Industry 4.0 is a wide view with clean constructions and credential plans, in a general sense depicted by the getting over of real present-day assets and automated developments in assumed computerized genuine structures [11]. One certain piece of Industry 4.0 is the value creation impacts from gains in adequacy and new game plans, nonetheless, inventive change may have both a positive and an antagonistic outcome on business [12]. Figure 8.5 shows Industry 4.0.

8.8 History of Industry 4.0

- First Industrial Revolution
- Second Industrial Revolution
- Third Industrial Revolution
- Fourth Industrial Revolution

8.8.1 First Industrial Revolution (Industry 1.0)

The First Industrial Revolution occurred in the eighteenth century with the use of steam power and robotization in production. What before made strings on undeviating exchanging wheels, the motorized design achieved it via various stages in a tantamount amount of time. Steam energy was by then known. Its usage for industrial goals was an awesome leap for extending man proficiency. Perhaps than material drift obliged by strength, steam engines could be exploited for competency. Upgrades such as the steamship or (almost 100 years sometime later) the vapor-administered train also accomplished an enormous upgrade in commutating many more people and items in fewer hours.

8.8.2 Second Industrial Revolution (Industry 2.0)

The Second Industrial Revolution was initiated in the 19th century, all around the openness of power and consecutive evolution of structure formation. Henry Ford (1863–1947) implemented the assembly line technique from an abattoir in Chicago: The pigs swung from transmission bands and each flesher carried out only a section of the task of devastating the animal. Henry Ford carried on with these postulates into auto-formation and intrinsically transmuted them simultaneously. Already one halt amassed a whole auto, as of then, the automobiles were made in partial steps on the vehicle line – through and through speedier and at a lesser price.

8.8.3 Third Industrial Revolution (Industry 3.0)

The Third Industrial Revolution set about during the 1970s in the 20th century across incomplete motorization employing memory-programmable controls and PCs. In the consideration of these headways, we were right then assembled to brutalize a whole formation measure in the absence of human cooperation. Familiar instances of this are robots that execute redid groupings in the absence of the mediation of man.

8.8.4 Fourth Industrial Revolution (Industry 4.0)

We are right now completing the Fourth Industrial Revolution. This is depicted by the utilization of data and correlation promotions to industry and is basically called "Industry 4.0." It evolves the improvements of the Third Industrial Revolution. Formation frameworks that as of now have PC advancement stretched out by a consortium adding and have an electronic twin on the Internet, figuratively speaking. These award relationships with various establishments and abdication of details regarding themselves. This is the ensuing level in advanced automation. The structure association of all frames gives rise to "advanced genuine creation systems" and thusly sharp industrial offices, in which formation frameworks, fragments, and people passing on utilizing an affiliation and formation is basically self-regulating.

Exactly when these enabling collisions gather, Industry 4.0 can move on a couple of mind-blowing progress under progress-line conditions. Models consolidate devices that can anticipate dissatisfactions and activate upkeep estimates self-rulingly or self-worked with coordinated efforts that react to frightening switches in progression.

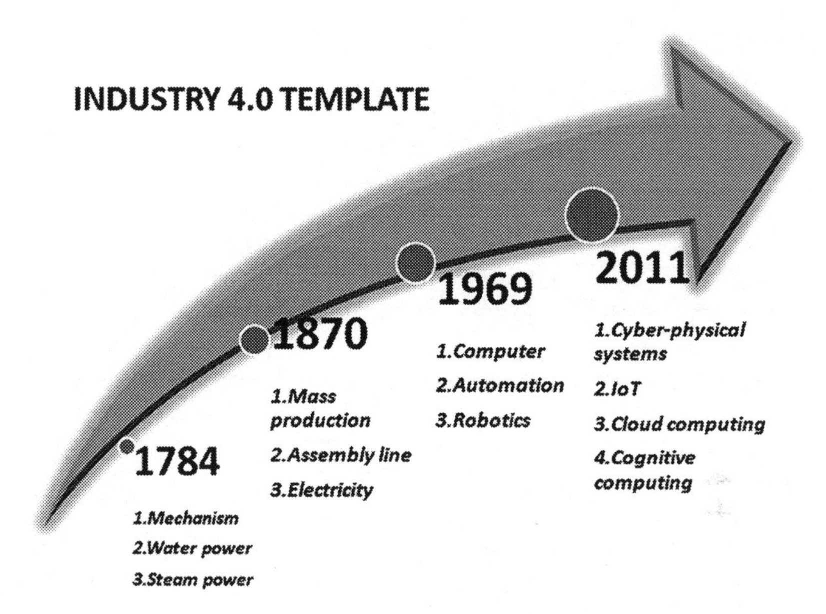

FIGURE 8.6
Industry 4.0 template.

Additionally, it can modify the way that people administer. Industry 4.0 can move people into more clever relationships, with the capacity of much helpful functioning. The digitalization of the social event climate thinks about more flexible techniques for acquiring the right data for the best individual at the best time. The developing implementation of state-of-the-art equipment interior creation lines determines furthermore, somewhere else in the field, if an upkeep expert can be given gear attestation and association history in a classic way, and at the spot of use. Upkeep experts are expected to oversee matters, not playing and attempting to begin with the specific information which they require [13].

Thus, Industry 4.0 is an unquestionable advantage over industrial surroundings. The digitalization of social occasions will modify how items are built and dispersed, and how items are upgraded and clarified. On that note, it can really advance a location resisting the inception of the Fourth Industrial Revolution. Figure 8.6 shows the Industry 4.0 template.

8.9 Pros and Cons of Industry 4.0

Here, we separate the advantages and disadvantages of Industry 4.0 to give you an understanding of what makes and mechanical affiliations to remember when moving toward advanced change drives [14].

8.9.1 Pros

Some of the advantages of Industry 4.0 are:

Competitive advantages

Expansion in operational efficiency

Better products and services

Development of markets and new markets

Further developing lives overall

- Competitive Advantages
  Industry 4.0 smart arrangements and administrations provide a broad scope of dominance for associations that can effectively dispatch these new methodologies and innovations.
- Expansion in Operational Efficiency
  The expectation of Industry 4.0 is that the new-to-the-scene age of modern unrest will drive considerably more noteworthy productivity for associations, as they can press more prominent yield from a similar asset input.
- Better Products and Services
  Whether it is a component, security, or client encounter, Industry 4.0 will work more on providing irrefutable and unmistakable components and throughputs, and thus allows to continue drawing in clients.
- Development of Markets and New Markets
  With any innovative unrest, new organizations, things, and programming will be expected to help the change of associations. This will make new item classes, and new positions, and that's just the beginning.
- Further Developing Lives Overall
  With new advancements, higher productivity, and development in economies, people groups' lives all in all, by and large improve, with pay rise, better wellbeing arrangements, and, in general, more prominence of life.

8.9.2 Cons

Here are some disadvantages of Industry 4.0:

Significant expenses

High rate of failure

Network protection

Need for highly skilled labor

Industry and market disruption

- Significant Expenses
  Not simply is development a huge expense to examine, in any case, the limit is in engaging the development to be completed. Having the aptitude in more current fields like IoT, Augmented Reality (AR), and AI can incite huge disbursing constraints and, additionally, a shortfall of understanding among all get-togethers included.
- High Rate of Failure
  The difficulty in forwarding Industry 4.0 is that there is reliably a shortage of hanging on for regards to creating targets. They are consistently cross-commonsense exercises with various accomplices, which can mean endeavors can become covered in conflicting destinations, and may simply vacillate out.
- Network Protection
  Individuals, things, and stuff is, and will continuously be, related with the web. Yet, this gives us more important induction to data through the cloud, and it opens up promising conditions for designers to form associations.
- Need for Highly Accomplished Labor
  Mass production, and assiduity in general, keeps on relying upon people to authorize formation. Nonetheless, with the change to painstakingly related structures, there is an increased conspicuous prerequisite for significantly accomplished duties, which may incidentally diminish the necessity for a low-mastery job.
- Industry and Retail Disturbance
  With current advances available, managing game plans will eventually be killed. Like the megahit of the world, unmistakable undertakings cannot endure what Industry 4.0 brings to the fore.

8.10 IIoT in SMEs

SMEs are endeavors that are ordered based on the size of speculation and they exist in both the Manufacturing and Services Sectors of the economy. In an enormous demography like India, SMEs give colossal monetary development, adding to about 24% of Services GDP and 6% of manufacturing GDP. Other than that, they likewise add to about 45% of net fares and give enormous work openings. The maintenance of development in SMEs is of public significance as they contribute altogether to the by-and-large monetary advancement of the economy and produces mammoth work in the country. Although much has been done effectively through facilitating limitations and guidelines, the interaction of enrollment should be made smoother and more problem-free. The World Wide Web (WWW) offers invigorating new freedoms

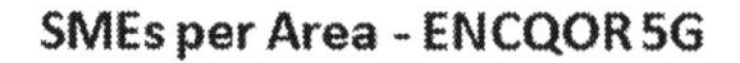

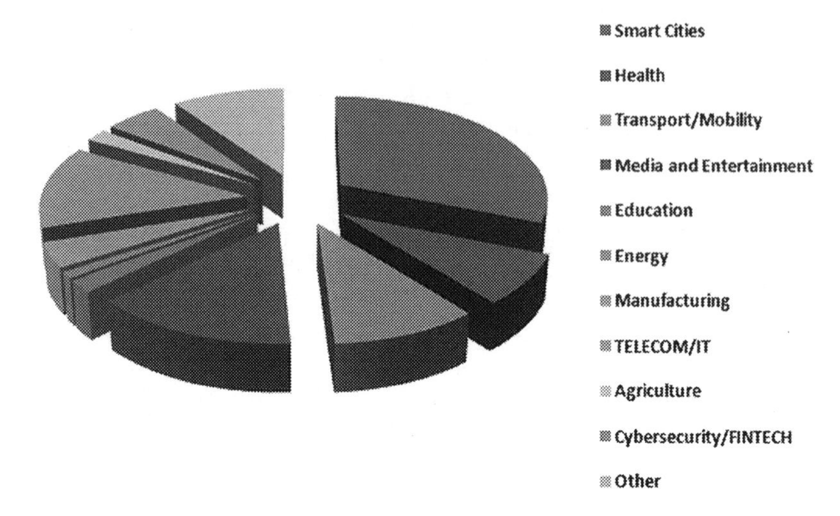

FIGURE 8.7
SMEs per area.

for SMEs to expand their client foundation into the worldwide commercial center [15]. Guidelines should be facilitated further and more credit openings are required around here to construct another age for venture and industrialization in the country. The very required boundary of SMEs emerged from the need for recognizable proof and advancement of these endeavors. With their appropriate recognizable proof and the production of a legitimate public information base of such ventures, the offices for the arrangement of advances customized to the necessities and prerequisites of this area could be handily cutting-edge without the issue of spillage. Accordingly, this has given credit firms the capacity to make advance headways without stress [16]. The organization of functional IoT arrangements in enormous organizations is around 20%, and in certain businesses, as high as 80%. However, in SMEs, reception is nearer to 10% with low degrees of general mindfulness revealed [17]. IIoT innovations are key for SMEs to carefully change their modern tasks and stay serious. IIoT reception is as yet an intricate and expensive interaction for SMEs, and worries about security stay a huge obstruction. Figure 8.7 shows the SMEs per area.

8.11 Smart Energy Manufacturing Systems

Smart manufacturing is a general characterization of gathering that uses PC-formed gatherings, huge stages of limberness and quick arrangement swaps, progressed information advancements, and added adaptable specific liveware getting ready. Various targets occasionally recall fast changes for creation levels reliant upon demand, headway of the stock organization, capable creation, and recyclability. In this thought, the smart industrial office has compatible frameworks, multi-scale vital exhibiting and reenactment, watchful computerization, strong advanced reliability, and orchestrated actuators. Levels of interest in smart manufacturing have been rising quickly – the greater part of producers have put essentially $100 million in the action [18]. The wide meaning of smart producing covers various advances. A portion of the vital innovations in the brilliant assembling development incorporates large information preparing capacities, mechanical network gadgets and benefits, and progressed mechanical technology. A smart energy management system is a PC-based framework intended to screen, control, measure, and streamline energy utilization in a structure, manufacturing plant, or any office. The frameworks can interface power devouring frameworks, like Heating, Ventilation, and Air Conditioning (HVAC), lighting, and assembling hardware, with meters, sensors, and different gadgets that can track measure, and total the information inside, and possibly speak with the utility or lattice administrator to time cooperations, for example, energy buys during less-expensive, off-top hours; investment sought after reaction occasions; EV oversaw charging projects; and transmission of surplus energy as a component of other utility or network administrator programs [19]. Smart Manufacturing is considered another worldview that makes work smarter and more associated, bringing velocity and adaptability through the introduction of modernized advancement. Today, advanced development is firmly connected to the "maintainability" of organizations. Advanced development and maintainability are two indistinguishable standards that depend on the idea of a round economy. Computerized development empowers the roundabout economy model advancing the utilization of arrangements like advanced stages, smart gadgets, and man-made consciousness that assist to upgrade assets. SMSs endeavor to expand those capacities by utilizing cutting-edge innovations that advance quick stream and inescapable utilization of computerized data inside and between fabricating frameworks [20].

8.12 Smart Grid

A Smart Grid is a high-level advanced two-way, stream power framework fit for self-recuperating, versatile, tough, and reasonable with prescience for expectation under various vulnerabilities [21]. It is a power network that can cost-productively coordinate the direct and exercises of all customers related to it – generators, purchasers, and those that do both – to ensure a fiscally successful, sensible power system with low adversities and huge levels of significant worth and security of supply and prosperity. Figure 8.8 shows the smart grid.

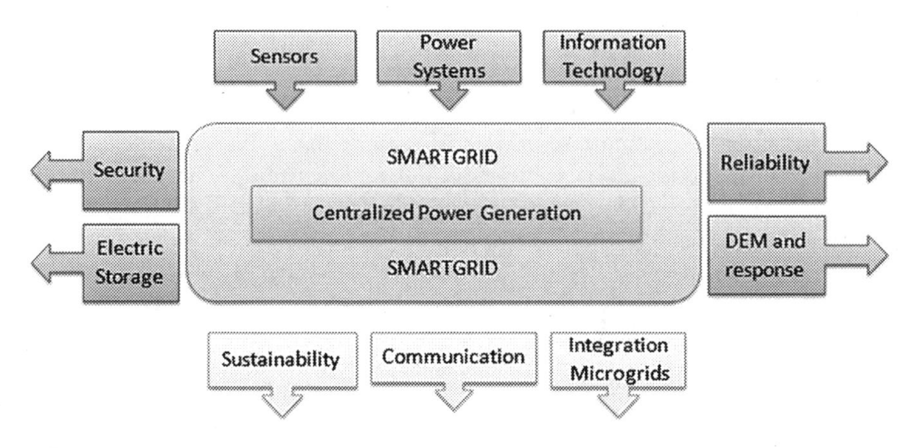

FIGURE 8.8
Smart grid.

A smart grid uses creative things and organizations alongside quick noticing, control, correspondence, and self-patching advances to [22]:

- Better work with the affiliation and action of generators, things being what they are, and advances.
- Allows purchasers to impact redesigning the movement of the framework.
- Furnish buyers with more noteworthy data and choices of how they utilize their inventory.
- Fundamentally diminish the ecological effect of the entire power supply framework.
- Keep up with or even work on the current undeniable degrees of framework dependability, quality, and security of supply.
- Keep up with and further develop current administrations productively.

8.13 Microgrid

A microgrid is a free power framework that sets out a discrete geographic impression like a school ground, clinic complex, business focus, or neighborhood. A superior method to understand the arising capability of disseminated age is adopting a framework strategy that sees age and related burdens as a subsystem or a "microgrid" [23]. Inside microgrids are at least one sort of conveyed energy (sun-based boards, wind turbines, consolidated warmth and force, generators, etc.) that produces its force. Likewise, numerous more current microgrids hold energy stockpiling, ordinarily, from batteries. Some additionally now have electric automobile charging stations. Interconnected to close structures, the microgrid gives power and conceivable warmth and cooling to its clients, conveyed through modern programming and control frameworks.

A microgrid is characterized by three lead attributes:

- A microgrid is local
- A microgrid is independent
- A microgrid is intelligent

Figure 8.9 shows the microgrid.

- Microgrid Is Local
  In the first place, this is a type of local energy, which means it makes energy for close-by clients. This recognizes microgrids from the sort of huge unified networks that have given a large portion of our power in the last century. Focal lattices push the power from generating stations over significant interspaces utilizing conveyance and dissemination lines. Conveying power from far off is wasteful because a portion of the power – approximately 8–15%) – disperses on the way. A microgrid defeats this shortcoming by producing

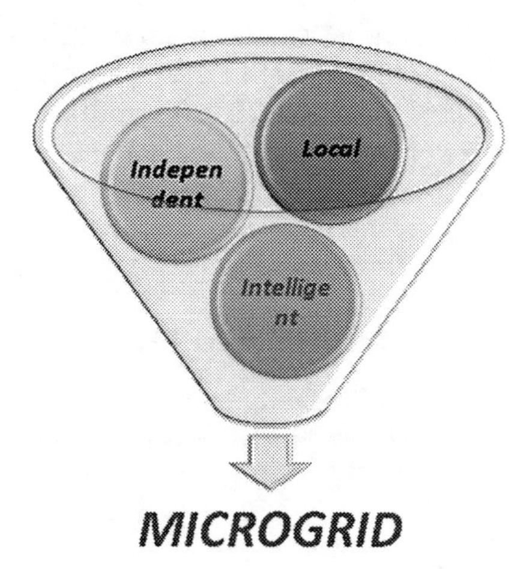

FIGURE 8.9
Microgrid.

power near those it serves; the generators are close to or inside the structure, or, on account of sunlight-based boards, on the rooftop.

- A Microgrid Is Independent
  Second, a microgrid can detach from the focal matrix and work alone. This sequester ability permits them to supply capacity to their clients when a tempest or other disaster causes a blackout on the force lattice. In the US, the focal network is particularly inclined to blackouts in light of its utter size and correspondence – more than 5.7 million miles of communication and dispersion lines. As we adapted agonizingly during what's known as the Northeast Blackout of 2003, a solitary tree tumbling on an electrical cable can take out power in a few states, even across worldwide limits into Canada. By islanding, a microgrid gets away from such falling lattice disappointments. While microgrids can run freely, more often than not, they don't (except if they are situated in a far-off region where there is no focal framework or a temperamental one). All things considered, microgrids commonly stay associated with the focal matrix.
- A Microgrid Is Intelligent
  Third, a microgrid – particularly progressed frameworks – is intelligent. This insight radiates based on what's known as the microgrid regulator, the focal mind of the framework, which deals with the generators, batteries, and close-by raising energy frameworks with a serious level of complexity. The regulator organizes different assets to encounter the energy objectives set up by the microgrid's clients. They might be attempting to accomplish the most minimal costs, cleanest energy, most noteworthy electric dependability, or some other result. The regulator accomplishes these objectives by expanding or diminishing utilization of any of the microgrid's assets – or blends of those assets – much as a supervisor would avail oneself of different artists to uplift, underneath, or quit taking part in their contraption for the greatest impact.

8.14 Impact of Industry 4.0 on Smart Manufacturing

Industry 4.0 is a task in the cutting-edge methodology of the German government that advances the mechanization of customary enterprises like assembling. The objective is the insightful production line (Smart Factory) that is portrayed by versatility, asset effectiveness, and ergonomics, just as the coordination of clients and colleagues in business and worth cycles. Its innovative establishment comprises digital actual frameworks and the IoT [24].

This sort of "intelligent manufacturing" utilizes:

- Remote associations, both during item gathering and significant distance connections with them
- Last age sensors, dispersed along with the store network and similar items (IoT)
- Expansion of a lot of information to rule all periods of development, circulation, and use of a decent.
- Progressed fabricating cycles and fast prototyping will specially make it feasible for every client stand-out item without huge expense increment.
- Cooperative Virtual Factory (VF) stages will radically lessen cost and time related to new item plans and designing of the creation cycle, by taking advantage of complete reproduction and virtual testing all through the product wheel of life.
- Progressed HMI and AR gadgets will assist to expand security underway plants and diminish actual interest in laborers (whose age has an expanding pattern).
- AI will be essential to advance the creation measures, both for lessening arrangement and decreasing energy utilization.
- Digital actual frameworks and M2M correspondence will permit assembly and transfer constant information from the shop floor to decrease vacation and inactive time by leading to incredibly viable prescient support.

8.15 Advantages and Disadvantages of Smart Manufacturing

8.15.1 Advantages

There are many benefits of assembling innovation that effectively assists with boosting creation, eventually saving the creation office cash and ensuring that creation is on time and capable [25].

- Expansion in Quality
  The quality redesign is by far one of the primary invaluable sections of collecting advancement. With creation programming, individuals are required less in all pieces of creation organizing and booking, similar to the genuine creation measure itself. Computerization in the creation of schedules and creation lines suggests a high-level plan that diminishes the number of disappointments, forsakes,

and various mishaps. This is on the grounds that individuals are more disposed to bungle than tweaked machines are, so it is clear why various creation workplaces are choosing to use robots and robotization instead of having endless number of workers inside the plant.

- Cost Reduction
  Cost decline is one of the basic targets of gathering advancement. This is an immediate consequence of the amendment of deficiencies and waste being decreased inside the creation correspondence, which saves an extreme proportion of money as time goes on. Assembling advances work on general usefulness, which builds benefit tremendously too. What's more, innovation and mechanization generally imply that you require fewer laborers in the plant, which is commonly the biggest expense brought about by an assembling organization.
- Decrease in General Manufacturing Time
  The lengthier the creation cycle, the higher the price. Gathering progressions operate on inventive collaborations and get things out in a considerably more useful manner. This is all appreciation of machines robotizing the connection, where creation time is diminished between thing clusters, finally thinking about the gathering action to grow benefits. Additionally, using machines to mechanize the creation cycle infers that you have a solid run rate for creation that can be used to all more definitely expect when you can pass on your product.
- Information-Driven Forecasting and Planning
  Advances in AI and creation arranging and booking implies that you will want to coordinate with your production network with genuine client requests all the more precisely. This will permit numerous makers to just deliver what is expected to supply the client's interest and try not to make squander because of overproduction driven by helpless anticipating.

8.15.2 Disadvantages

Alongside the different benefits of assembling innovation, there are hindrances relating to the innovation too. The disservices of assembling innovation incorporate the following.

- Restricted Creativity
  Assembling innovation confines creativity on account of the abundance of robotization/equipment and the shortfall of agents inside the creation office. Representatives can conceptualize while

balancing a specific issue, while the hardware is just upgraded to get in line, regardless of whether there are issues.

- High Initial Cost
  Many makers are charmed by the expense decrease advantages of executing robotization and innovation inside their assembling office. While the advantages can assist with reducing expenses and eventually increment benefits, there can be high beginning speculation costs included. Evaluate well the long benefits of adding different kinds of innovation to your tasks to decide if the fundamental execution cost will be wonderful.
- Commitment to Environmental Issues
  An Earth-wide temperature help is a tremendous concern for certain individuals throughout the planet – with assembling being a liberal ally of it. Assembling innovation implies greater hardware and innovation being fused with creation offices, which contrarily affects the climate because of the different fuel sources, for example, gasses and synthetics needed to run it. What's more, a more effective creation line implies that you can build your overall creation aggregate – prompting more waste and different discharges.
- Joblessness Increase
  Joblessness has been a gigantic concern for gathering since robotization began to accept a section. Since the time when making movements have become typical, there has been a great deal of stress over how that influences human work inside collecting workplaces. For some, adding more machines and mechanized cycles implies that less work is required. Then again, a greater amount of your workers' time would now have to be able to be committed to high-esteem errands that will assist with developing your business.

8.16 Conclusion

In conclusion, the IoT is the thought where the effective cosmos of data improvement is connected with this current actuality of things. The IoT is somehow a primary way to the shrewd world with certain figuring and structures' association to ease different endeavors around customers and give various tasks, for instance, basic checking of different wonders incorporating us. In the IoT, normal things from regular day-to-day existence, named "things" or "machines" are redesigned with preparation and correspondence advancements. Industry 4.0 is provoking some obviously perceptible and quantifiable benefits, for instance, lower costs, more capability, easier stock organization, lower compensation time, and improved usefulness. Industrial IoT can fundamentally change production. Truly, an ever-increasing quantity of information will be made from progressively associated hardware frameworks and acknowledging basic and important significant bits of knowledge through IIoT is boundless. In order to learn through IIoT-based Smart Manufacturing in SMEs, the contents of IoT, IIot, SMEs, and smart manufacturing are discussed together with their pros and cons.

References

1. Golpîra, Hêriş, Syed Abdul Rehman Khan, and Sina Safaeipour. "A review of logistics internet-of-things: Current trends and scope for future research." *Journal of Industrial Information Integration* 22 (2021): 100194.
2. Sadeeq, Mohammed AM, and Subhi Zeebaree. "Energy management for internet of things via distributed systems." *Journal of Applied Science and Technology Trends* 2.02 (2021): 59–71.
3. Wang, Jianxin, et al. "The evolution of the Internet of Things (IoT) over the past 20 years." *Computers & Industrial Engineering* 155 (2021): 107174.
4. Lv, Zhihan, et al. "Big data analytics for 6G-enabled massive internet of things." *IEEE Internet of Things Journal* 8.7 (2021): 5350–5359.
5. Mezzanotte, Paolo, et al. "Innovative RFID sensors for Internet of Things applications." *IEEE Journal of Microwaves* 1.1 (2021): 55–65.
6. Halder, Subir, and Thomas Newe. "Enabling secure time-series data sharing via homomorphic encryption in cloud-assisted IIoT." *Future Generation Computer Systems* 133 (2022): 351–363.
7. Yu, Keping, et al. "Blockchain-enhanced data sharing with traceable and direct revocation in IIoT." *IEEE Transactions on Industrial Informatics* 17.11 (2021): 7669–7678.
8. Cakir, Mustafa, Mehmet Ali Guvenc, and Selcuk Mistikoglu. "The experimental application of popular machine learning algorithms on predictive maintenance and the design of IIoT based condition monitoring system." *Computers & Industrial Engineering* 151 (2021): 106948.
9. Wang, Oliver. "Connectivity enables smart production lines: Before reaping the benefits of IIoT, operators need to gain visibility of the production floor." *Plant Engineering* 76.1 (2022): 37–40.
10. Zheng, Ting, et al. "The applications of Industry 4.0 technologies in manufacturing context: A systematic literature review." *International Journal of Production Research* 59.6 (2021): 1922–1954.
11. Acioli, Carina, Annibal Scavarda, and Augusto Reis. "Applying Industry 4.0 technologies in the COVID–19 sustainable chains." *International Journal of Productivity and Performance Management* (2021).
12. Nara, Elpidio Oscar Benitez, et al. "Expected impact of industry 4.0 technologies on sustainable development: A study in the context of Brazil's plastic industry." *Sustainable Production and Consumption* 25 (2021): 102–122.

13. Chehri, Abdellah, et al. "Theory and practice of implementing a successful enterprise IoT strategy in the industry 4.0 era." *Procedia Computer Science* 192 (2021): 4609–4618.
14. Verma, Anju, and M. Venkatesan. "Industry 4.0 workforce implications and strategies for organisational effectiveness in Indian automotive industry: a review." *Technology Analysis & Strategic Management* (2021): 1–9.
15. Ying, Zhen, et al. "An overview of computational models for industrial internet of things to enhance usability." *Complexity* (2021).
16. Sharma, Durga Prasad, et al. "Emerging paradigms and practices in cloud resource management." *Autonomic Computing in Cloud Resource Management in Industry 4.0*. Springer, Cham, 2021. 17–39.
17. Indoria, Mayank, et al. "Implementation of industry 4.0 to achieve sustainable manufacturing in steel industry: A case study." *Systematic Literature Review and Meta-Analysis Journal* 2.1 (2021): 1–9.
18. Zuo, Yanjun. "Making smart manufacturing smarter–a survey on blockchain technology in Industry 4.0." *Enterprise Information Systems* 15.10 (2021): 1323–1353.
19. Cots, Ainhoa, et al. "Energy efficient smart plasmochromic windows: properties, manufacturing and integration in insulating glazing." *Nano Energy* 84 (2021): 105894.
20. Bermeo-Ayerbe, Miguel Angel, Carlos Ocampo-Martinez, and Javier Diaz-Rozo. "Data-driven energy prediction modeling for both energy efficiency and maintenance in smart manufacturing systems." *Energy* 238 (2022): 121691.
21. Krishnan, Priya R., and Josephkutty Jacob. "An IOT based efficient energy management in smart grid using DHOCSA technique." *Sustainable Cities and Society* 79 (2022): 103727.
22. Saleem, M. Usman, et al. "Design, deployment and performance evaluation of an IoT based smart energy management system for demand side management in smart grid." *IEEE Access* 10 (2022): 15261–15278.
23. Sedhom, Bishoy E., et al. "IoT-based optimal demand side management and control scheme for smart microgrid." *International Journal of Electrical Power & Energy Systems* 127 (2021): 106674.
24. Gerekli, İsa, Tarık Ziyad Çelik, and İbrahim Bozkurt. "Industry 4.0 and smart production." *TEM Journal* 10.2 (2021): 799–805.
25. Kotsiopoulos, Thanasis, et al. "Machine learning and deep learning in smart manufacturing: The smart grid paradigm." *Computer Science Review* 40 (2021): 100341.

9

The Role of Cybersecurity in Smart Cities

Azeem Khan and Noor Zaman Jhanjhi
Taylor's University, Subang Jaya, Selangor, Malaysia

Mamoona Humayun
Jouf University, Al Jouf, Saudi Arabia

CONTENTS

9.1 Introduction

Cities have been centers of civilization, symbolizing human progress ever since the beginning; due to this individuals and masses have been migrating toward cities in search of the betterment of living and this process is still in continuum and will be in the continuum due to opportunities that make human lives better in terms of economic and professional growth. The rural economy is primarily agrarian; hence, after the process of industrialization

DOI: 10.1201/9781003203087-9

had begun, it paved the way for mass production, creating huge employment opportunities in cities; henceforth, people started flooding toward cities globally increasing the population in cities. With the advent of Information and Communication Technologies (ICTs), the process of automation started, which is still in continuum [1–2]. The application of ICT technologies in cities provisioned all the stakeholders involved in this ecosystem in tackling the problems associated with the administration of cities pertaining to basic infrastructures such as mobility, health, housing, education, public safety, environment, and energy thus paving the way for the emergence of cities that are interconnected and embedded with intelligent devices and were termed as smart cities. Smart cities are cities that are surrounded by a plethora of devices [3–5] that are interconnected and intelligent. These technologies have eased the local governments in terms of increased efficiency in administration by improvising the basic facilities required by the inhabitants of the cities. Though civilian lives have been improved a lot and contributed to an enhanced quality of life, they have created many other challenges that are at the cost of compromised security and privacy. To address these challenges and to provide better solutions, cybersecurity has been proven to be of utmost significance and can't be undermined. To simplify this concept of smart cities, we have summarized it with conceptual layers and the risks associated with them, as illustrated in Figure 9.1.

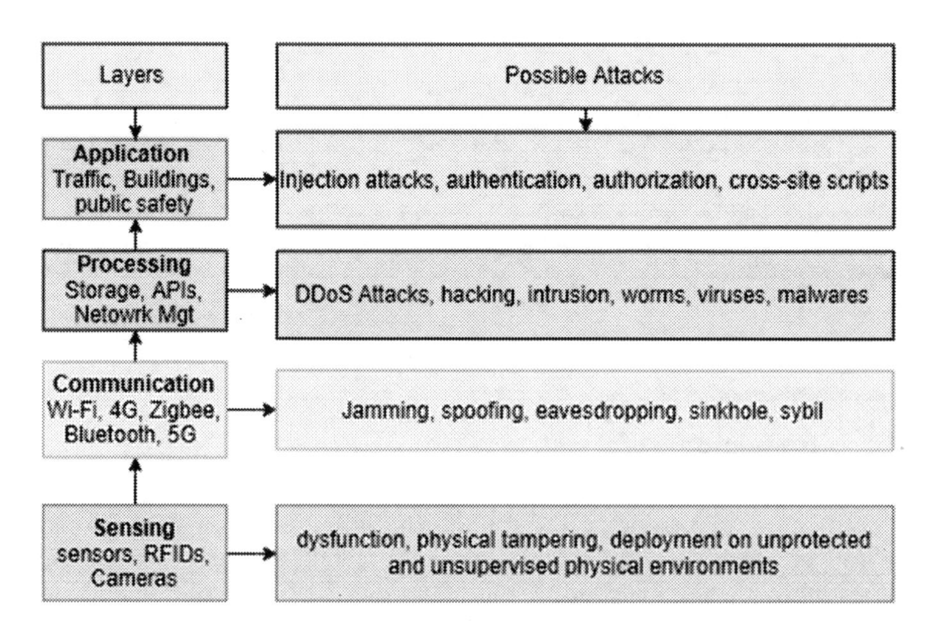

FIGURE 9.1
Cybersecurity attacks on various smart city infrastructure layers.

The smart city architecture is envisioned as four layers. Each layer is connected with a distinct set of technologies and numerous possible cyberattacks associated with them, which can't be undermined at any cost; these are labeled above concerning each layer. Thus, succinctly, smart city provides digital infrastructure comprising ubiquitous sensing enabled through sensors, and communication among these is enabled with communication technologies such as WSN, 4G, Zigbee, Bluetooth, Z-wave, Near Field Communication (NFC), 5G, etc., and processing of the enormous data generated by this plethora of connected devices, which is termed as Big Data, is facilitated with cloud computing technology through data analytics by implementing sophisticated machine learning algorithms which are helpful in providing meaningful insights with the existing data generated on a real-time basis. This has empowered all the participants of this smart city environment by availing them to take intelligent decisions on time and improving the lives of citizens dwelling in cities who are able to access all the smart city infrastructure via applications in the form of apps [6–8].

Each layer of this infrastructure is associated with the risks as well; at the application layer, the users' systems can be compromised with cyberattacks that may be cross-site scripts or injection attacks; apart from it, a common trap is gaining access to the system through authentication and authorization; there are also many other security risks at various layers, such as intrusion of viruses, worms, jamming, DDoS attacks at processing and communication layer, and last but not least, at the sensing layers, possible attacks and physical tampering of the devices typically leading to malfunction if they are deployed on unprotected and unsupervised physical environments [9].

The remaining part of the chapter is arranged as follows. Section 9.2 throws light on the emerging state-of-the-art technologies for smart cities; Section 9.3 is focused on expounding various applications and their cybersecurity threats associated with them; Sections 9.4 and 9.5 are dedicated to elaborating the proposed framework, open issues, and challenges pertaining to the cybersecurity of smart cities. Lastly, it is concluded with a note on possible future works.

9.2 Emerging State-of-the-Art Technologies for Smart Cities

There are several technologies involved in maintaining managing and securing the whole smart city ecosystem; to name a few are the emerging state-of-the-art technologies such as cloud computing, Big Data, and the Internet of Things (acronymized as IoT) [10], and among the communication technologies are the technologies such as RFIDs, WSN, Bluetooth, NFC, Zigbee, Z-wave, 4G, and 5G. They are summarized as follows.

9.2.1 Cloud Computing

In smart cities, data are centralized so that all stakeholders can access it. Cloud computing serves this purpose by providing its services through data centers that may be managed by the public sector or private sector or a mixture of both entities known as hybrid clouds [11–12]. Apart from providing this service, cloud computing also provides services for hardware, infrastructure, software, and platform as well, thereby saving cost and facilitating the fastest delivery of data with reliability, scalability, and high performance for large-scale computing tasks. Thus, cloud computing provides virtual infrastructure for devices with which smart cities can store, monitor, analyze, and visualize the data. However, cloud computing in a smart city isn't only about making the city more efficient. Data may be kept, evaluated, and used by administrators and governments in the short term to take the appropriate measures. However, in the long run, specific data and data-driven insights from one cloud might be sold to cloud providers in other cities, eventually leading to the construction of a "template" management system that could swiftly turn a "regular" city into a smart city with less human participation [13–15].

9.2.2 Big Data

The data are pivotal and have paramount significance in this whole ecosystem of smart cities. Data created by digital technology such as mobile phones, the internet, satellites, or sensors are referred to as Big Data. Data science – or artificial intelligence – technologies or methodologies are required to efficiently gather, curate, organize, and analyze these data. Big Data is frequently a by-product of our digital behavior, based on the digital "crumbs" we leave behind as we interact with systems or devices in our everyday lives – the enormous amount of data that is being generated by connected devices enabled with communication between them [16]. Data are collected through various data sources such as sensors, computers, cameras, Global Positioning Systems (GPSs), social networking sites, commercial transactions, and games; if this is processed and appropriately refined with apt state-of-the-art technologies, it can provide valuable insights with the application of Big Data analytics which run on sophisticated machine learning algorithms enabled with Artificial intelligence helps in effective and efficient administration of cities [17] with better traffic management in mobility by providing an insight with the real-time traffic data which help the dwellers of the cities make apt decisions for their travel. Hence, Big Data plays a vital role in availing this opportunity for the inhabitants of smart cities [18].

9.2.3 IoT

At the crux of the smart city ecosystem lies devices known as sensors which sense the data from the environment through which they are connected via the internet and are responsible for creating a massive amount of data are termed as objects that are interconnected and can communicate with each other as the IoT [18]. IoT devices are responsible for generating real-time data that are then cloud-stored at a centralized location. They are available for all the stakeholders connected to this seamless plethora of networks availing them to take intelligent decisions [19]. These devices sense temperature, humidity, noise levels, traffic congestion, public safety, signaling, etc. Although IoT devices play a vital role in smart cities, they are also highly vulnerable to security attacks. Therefore, vendors, device makers, and governments must work together to establish more strict IoT security regulations. Organizations and device makers must follow developing standards and advice to guarantee that systems are "secure by design" and that faults are identified and addressed before and after installation. Further, operators of active smart city technologies must also attempt to comprehend the security challenges that their smart surroundings and systems face if they are to minimize risks before accidents occur. Cities of the future will become smarter as time passes. Still, the distinction between a smart city and a safe city will be determined by how well IoT security is implemented.

9.2.4 RFID

Assets are tracked in smart cities with the help of Radio Frequency Identifiers, acronymized as RFID. They come in the form of tags and as patched-on objects, thereby providing all objects with a unique ID with which their whereabouts and assets inventory are made easy for the whole smart city landscape [20]. This technology operates on electromagnetic fields which can identify and track all the objects via tags, which are in rampant use in many areas associated with smart cities such as hospitals, schools, hotels, shopping malls, educational institutions, libraries, cargo, production assembly lines, etc., to name a few. Unlike barcode signals, RFID can be read even when it is not visible.

9.2.5 WSN

The acronym WSN stands for Wireless Sensor Networks and as the name suggests, it provides network connectivity without a physical connection. These networks are distributed and autonomous. WSNs are sensor nodes that measure temperature, light, asphyxiating gases/smoke, humidity, use of energy, and occupancy in intelligent building management systems. The communication between the devices connected through WSN is done through sensors. Sensors are low-power integrated circuits with wireless communication technology. They have proved to be ideal in several areas: natural disaster prevention, health monitoring, industrial process monitoring, home [19], water quality monitoring, etc. The characteristic feature of this network is it can monitor physical and environmental conditions in real-time such as humidity, pressure, light, and temperature, and are applicable for smart homes, smart buildings, and smart health [20].

9.2.6 4G and 5G

These communication technologies connote paradigm shifts in terms of data throughputs. 4G data speed ranges from 100 Mbps to1 Gbps. 4G can provide ultra-broadband internet service. 4G delivers the following services for the inhabitants of smart cities: communication through wearables, high-definition voice, short message service, multimedia message services, mobile TV, high-definition streaming, global roaming, gaming services, etc.; conversely, 5G is the latest communication technology that provides a platform for hundred billion devices and supports a bandwidth of 10 Gbps; 5G is capable of supporting virtual reality environment with ultra-HD audio/video applications along with 10 Gbps data speed for mobile cloud service, which is necessary for almost all the devices connected with the smart city ecosystem [21–22].

9.3 Applications for Smart Cities and Cybersecurity

For simplicity, we have categorized the applications of smart cities into various domains based on their utility. As depicted in Figure 9.2, they are broadly classified based on industry, energy, mobility, living services, and environment. Under mobility, smart cities support the basic infrastructure of transportation through smart transport systems, and to avoid traffic congestion, proper parking arrangements are made through smart parking [23–24]made available through RFID. These e-valets work on NFC technology enabling easy payment of billing. Smart traffic management is enabled with the sensors attached to the traffic signaling systems which are controlled and monitored continuously by traffic management authorities with the data being sent by the sensors installed at traffic signaling systems; these sensors sent continuous data updates through the cloud to the concerned authorities thereby assisting them to manage traffic through proper real-time data analytics [19].

Apart from it, there are other applications pertaining to generating energy, which is termed as smart energy wherein the sensors are attached

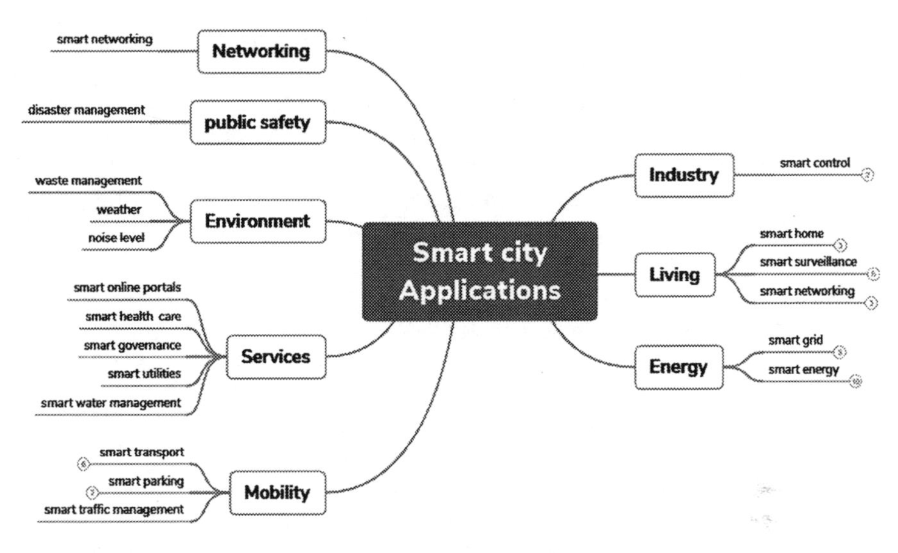

FIGURE 9.2
Smart city applications.

to windmills and solar panels which are the major source for producing renewable energy. Smart grids enabled with smart meters enable easy tracing of energy consumption, and smart street lighting facilitates smart city to optimize energy resources by using them whenever they are required and, at later times, switching them off through the concerned authorities who are able to access, monitor, and manage these resources remotely. Smart living avails smart home infrastructure embedded with smart surveillance [25], which is enabled with cameras that are installed at homes to monitor their inside and outside vicinities and make timely decisions in case of uncertainties and emergencies. Apart from it, there are numerous other applications such as public safety systems that are activated during the time of natural calamities by producing information earlier and facilitating it to ensure the safety of lives, and in times of theft, physical attacks by strangers can also be traced by these systems, enabling a better quality of living for city dwellers. State-of-the-art facilities are provided through services in health care, efficient governance improving quality of life through proper waste management, and controlling noise pollution, radiation levels, and green gas emissions through the data being generated by the sensors concerned with these applications. Apart from these, better employment opportunities are made available through smart industries embedded with cyber-physical systems that can monitor the health of machines connected within the industries. In this way, we have numerous applications that are connected to smart cities.

As discussed earlier, smart cities are equipped with sensors and actuators. Sensors are meant to collect data and pass them to the cloud, and actuators allow devices to act, for instance, by altering the streetlights in smart street lighting systems, restricting the flow of water the pipe with leakage, etc. By utilizing GPS data from the driver's smartphones, smart traffic management can be done more efficiently to understand; let us consider a scenario where the traffic jam can be prevented. To prevent traffic congestion in cities, the data retrieved from drivers are sent through the cloud to the traffic management team, which automatically sends a notification to the drivers, advising them to take an alternative route. Simultaneously, staff at a traffic control center are informed to send a congestion alert to relieve the congestion and reroute part of the traffic; they can send a command to the traffic light actuators to alter the signals and thereby solve the traffic congestion problem [23].

These services will grow increasingly vulnerable to cyber-attacks as smart cities become more linked and have more sophisticated digital infrastructure. Smart cities are just as powerful as their weakest link, and even the tiniest weakness may be exploited to tremendous effect. As a result, governments must pay greater attention to cybersecurity and spend more on threat prevention. Otherwise, governments would be constantly on the defensive as hackers exploit security flaws. The sources of these flaws, on the other hand, are well-known. IoT device security problems such as lack of proper encryption and patching over the wire are well known. Even though there is a lot of advice on how to make the devices and supporting architecture secure, the problems still persist. Manufacturers continue to push insecure products and systems off manufacturing lines into smart city solutions, expanding system attack surfaces, despite the lack of security standards [24–25].

All the above discussion put aside, the whole system can be sabotaged if these devices are compromised, hence there is a dire need to implement security measures to control any uneven happening in the entire ecosystem of smart cities. There have been incidents reported wherein the traffic signals were hacked and smart grids were compromised, smart surveillance systems were attacked, etc. Hence, we can't undermine the security intricacies at the cost of the ease these systems provide for the inhabitants of smart cities. We have proposed a framework for smart cities to address these cybersecurity issues, which is discussed in Section 9.4.

9.4 Proposed Framework of Cybersecurity and Smart Cities

As discussed above, smart cities have enormous potential benefits to serve the inhabitants of the cities; the other side, which is considered to be the darker side, is that they are vulnerable to attacks sabotaging the hardware and making the software services unavailable, software errors, unauthorized

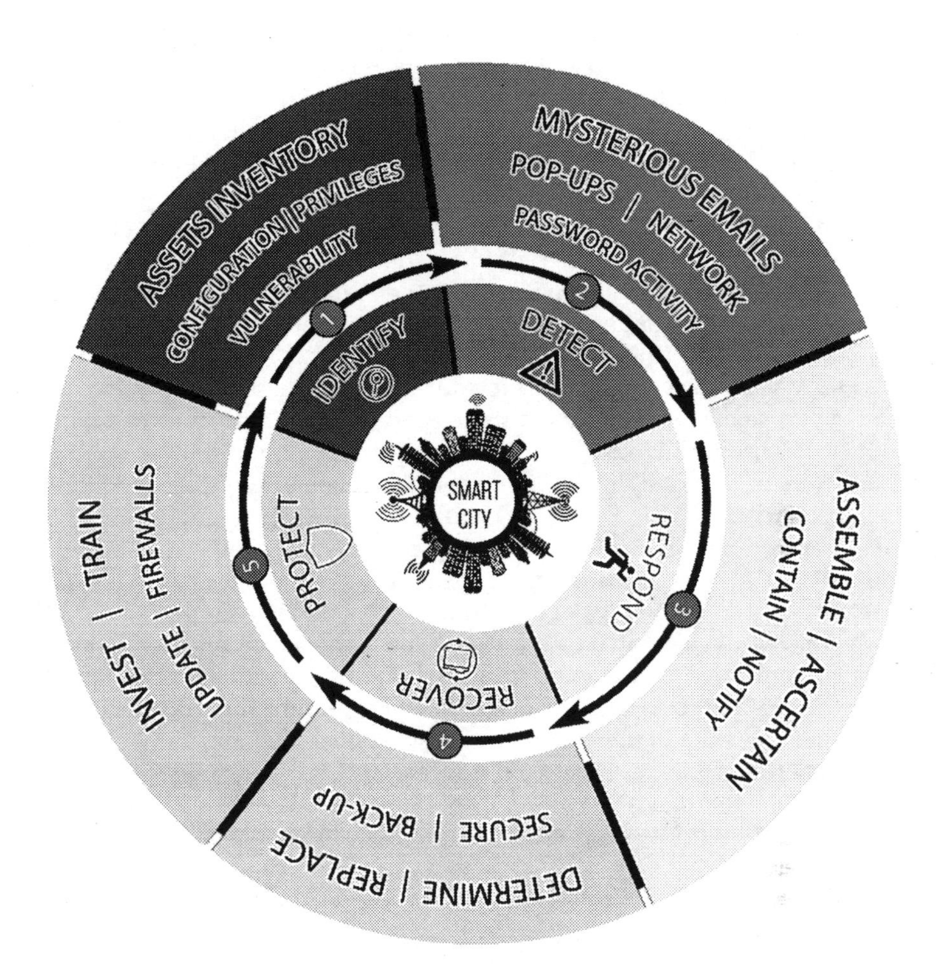

FIGURE 9.3
Proposed cybersecurity framework for smart cities.

usage or access to the system, data theft, malware operational or user errors, etc., to name a few. Hence, there is a need to tackle the issues pertaining to this darker side which are termed cybersecurity issues. To address these issues, we have proposed a framework which is depicted in Figure 9.3. The following are the steps that need to be implemented by all the stakeholders of this ecosystem.

The first and foremost thing every individual entity connected to this system needs to formulate is a strategy to overcome the threats inherent in this ecosystem. Our framework illustrates that first and foremost, individuals or organizations who are part of this smart city ecosystem need to identify their assets which can be hardware assets or software assets, and their inventory should be maintained continuously at regular intervals failing which intruders can easily creep in and sabotage the whole system with the weakest link. Apart from its regular configuration checks, the privileges allotted should also be reviewed frequently to comprehend the vulnerabilities creeping into the system. The second phase is to detect the vulnerabilities which can be done through regular monitoring of the network if any mysterious emails are found they need to be quarantined; unusual password activities should be taken with utmost significance and need to be checked; apart from that, suspicious popups and network slowdown are also symptoms that indicate the system is under attack, which should be addressed instantly and this can be done in the next stage which is termed as the respond stage which includes assembling experts teams to address relevant issues with respect to their expertise by ascertaining the source and by containing the susceptible entities; after doing all these, they should notify all the users of these systems to be aware of this kind of vulnerabilities incurring in the system and guide them for further actions.

The fourth phase is to recover by taking regular back-ups to determine and replace unwanted entities in the system. Last but not least is the phase is to protect the whole system by training the employees pertaining to this whole ecosystem; apart from it, the software pertaining to the operating system, anti-viruses, and firewalls should be updated regularly; however, this entails cost and hence the stakeholders should never step back from investing in them; otherwise, the whole system may be compromised, creating a havoc. To avoid hindrance and to avail smooth working of the whole smart city ecosystem, the aforesaid framework should be implemented without compromise; otherwise, the whole system may be compromised.

9.5 Open Issues and Challenges

Smart cities are comprehended through the involvement of several infrastructures assembled with embedded systems that are interconnected through heterogeneous network platforms with distributed systems. Since this whole ecosystem is involved with may many key players, it is proven to have weak links that pave the way for security and privacy breaches in this whole connected landscape and hence, smart cities are not an exception to security breaches and are vulnerable to several attacks which can be from internal entities or external or a combination of both. Smart cities are susceptible to various attacks comprising misconfiguration, malfunctioning, malicious misuse, botnets, ransomwares [26,27], campaigns, DDoS attacks, etc., and the major risk for the whole ecosystem is when they get compromised by the intruder [28].

Thus, it is a challenge to address these inherent issues related to smart cities. Smart cities' open issues and challenges can be summarized as follows: business challenges – these include costs incurred in implementing the cyber-physical systems with great investment; secondly, regulatory issues – proper regulations should be determined by the policy-makers [29] for the smooth functioning of the system. Sustainability highlights the elements that the whole ecosystem needs to achieve for these robust methods to be implemented for the system to sustain. Finally, interoperability is also a challenge that needs to be addressed as the whole ecosystem comprises different technologies as well as devices belonging to different domains; hence, there should be a proper mechanism that should be framed to address the operability among them and technical challenges pertaining to privacy that involves confidentiality, integrity, availability, and security concerns are related to unauthorized access [30].

9.6 Conclusion

From the preceding discussions, it is obvious that the smart city concept has revolutionized the way people think about urban living. The whole automated systems implemented with ICT have made urban living easier at the cost of technology which sometimes can create havoc if properly not managed; thus, concerned authorities should ensure that smart cities – which are instrumented, interconnected, and intelligent – are safer and more secure as well because of the many security and privacy concerns that are found to be loopholes in the whole system.

Thus, this chapter has been an effort to understand smart city is and its elements. We have proposed a framework that, if implemented properly by the concerned authorities, can make the whole smart city ecosystem work fine and run smoother. Not all vulnerabilities of smart cities are considered associated with hardware and software alone and there are, in fact, many instances where there were human errors. Therefore, it is critical and unambiguous to define roles specifically for those positioned in management,city administration, and cybersecurity officers.

Hence, a strong leadership to lead this whole ecosystem is needed and in its absence, there will be damages that cannot be recovered easily; henceforth, there is a dire need to implement strong verification checks, firewall updates, developing proper emergency response strategies and implementing proper authentication and authorization schemas that can be implemented using two-factor or three-factor authentication. The proposed cybersecurity framework facilitates achieving the whole smart city ecosystem's safety, security, and privacy.

9.7 Future Work

Though it seems a lot of work has been done in this area, in fact, there are many questions pertaining to security and privacy which are still unanswered and hence there is a dire need to explore more and provide better solutions for maintaining security and privacy of the whole smart city ecosystem.

The proposed research cybersecurity framework for smart cities may enable secure and efficient communication, apart from lending other transactions of smart city function safer, smoother, and more efficient in terms of its all-embracing functionality. It can be adopted by smart cities.

References

1. E. Bibri and J. Krogstie, "Smart sustainable cities of the future: An extensive interdisciplinary literature review," *Sustainable Cities and Society*, vol. 31, pp. 183–212, 2017.
2. M. Vitunskaite, Y. He, T. Brandstetter, and H. Janicke, "Smart cities and cyber security: Are we there yet? A comparative study on the role of standards, third party risk management and security ownership," *Computers & Security*, vol. 83, pp. 313–331, 2019.
3. Z. A. Almusaylim and N. Jhanjhi, "Comprehensive review: Privacy protection of user in location-aware services of mobile cloud computing," *Wireless Personal Communications*, pp. 1–24, 2019.
4. H. Habibzadeh, B. H. Nussbaum, F. Anjomshoa, B. Kantarci, and T. Soyata, "A survey on cybersecurity, data privacy, and policy issues in cyber-physical system deployments in smart cities," *Sustainable Cities and Society*, vol. 50, p. 101660, 2019.
5. P. I. Radoglou Grammatikis, P. G. Sarigiannidis, and I. D. Moscholios, "Securing the internet of things: Challenges, threats and solutions," *Internet of Things*, vol. 5, pp. 41–70, 2019.
6. M. I. Khalil, N. Z. Jhanjhi, M. Humayun, S. K. Sivanesan, M. Masud, and M. Shamim Hossain, "Hybrid smart grid with sustainable energy efficient resources for smart cities," *Sustainable Energy Technologies and Assessments*, vol. 46, p. 101211, 2021.
7. M. Humayun, N. Z. Jhanjhi, M. Z. Alamri, and A. Khan, "Smart Cities and Digital Governance," In *Employing Recent Technologies for Improved Digital Governance*, pp. 87–106. IGI Global, 2020.
8. B. Hamid, N. Z. Jhanjhi, M. Humayun, A. Khan, and A. Alsayat, "Cyber security issues and challenges for smart cities: A survey," In *2019 13th International Conference on Mathematics, Actuarial Science, Computer Science and Statistics (MACS)*, 2019: IEEE, pp. 1–7.

9. A. Fournaris, K. Lampropoulos, and O. Koufopavlou, "End Node Security and Trust vulnerabilities in the Smart City Infrastructure," *MATEC Web of Conferences*, vol. 188, 2018, p. 05005.
10. M. Aly, F. Khomh, M. Haoues, A. Quintero, and S. Yacout, "Enforcing security in internet of things frameworks: A systematic literature review," *Internet of Things*, vol. 6, 2019.
11. C. Stergiou, K. E. Psannis, B. B. Gupta, and Y. Ishibashi, "Security, privacy & efficiency of sustainable cloud computing for big data & IoT," *Sustainable Computing: Informatics and Systems*, vol. 19, pp. 174–184, 2018.
12. J. Sengupta, S. Ruj, and S. Das Bit, "A comprehensive survey on attacks, security issues and blockchain solutions for IoT and IIoT," *Journal of Network and Computer Applications*, vol. 149, p. 102481, 2020.
13. E. Al Nuaimi, H. Al Neyadi, N. Mohamed, and J. Al-Jaroodi, "Applications of big data to smart cities," *Journal of Internet Services and Applications*, vol. 6, no. 1, 2015.
14. I. A. T. Hashem et al., "The role of big data in smart city," *International Journal of Information Management*, vol. 36, no. 5, pp. 748–758, 2016.
15. F. A. Alaba, M. Othman, I. A. T. Hashem, and F. Alotaibi, "Internet of things security: A survey," *Journal of Network and Computer Applications*, vol. 88, pp. 10–28, 2017.
16. S. Alayda, N. A. Almowaysher, M. Humayun, and N. Z. Jhanjhi, "A novel hybrid approach for access control in cloud computing."
17. B. A. Alenizi, M. Humayun, and N. Z. Jhanjhi, "Security and privacy issues in cloud computing," In *Journal of Physics: Conference Series*, vol. 1979, no. 1, 2021: IOP Publishing, p. 012038.
18. M. Humayun, "Role of emerging IoT big data and cloud computing for real time application," *International Journal of Advanced Computer Science and Applications(IJACSA)*, vol. 11, no. 4, 2020. http://dx.doi.org/10.14569/IJACSA.2020.0110466
19. Z. A. Almusaylim and N. Zaman, "A review on smart home present state and challenges: Linked to context-awareness internet of things (IoT)," *Wireless Networks*, 2018.
20. M. Shafiq, H. Ashraf, A. Ullah, M. Masud, M. Azeem, N. Z. Jhanjhi, and M. Humayun, "Robust cluster-based routing protocol for IoT-assisted smart devices in WSN." *CMC-Computers Materials & Continua*, vol. 67, no. 3 (2021): 3505–3521.
21. J. Dwivedi et al., "RFID technology-based attendance management system," *International Journal of Engineering Science and Computing (IJESC)*, vol. 7, no. 3, 2017.
22. M. Humayun, N. Jhanjhi, M. Alruwaili, S. S. Amalathas, V. Balasubramanian, and B. Selvaraj, "Privacy protection and energy optimization for 5G-aided industrial internet of things," *IEEE Access*, vol. 8, pp. 183665–183677, 2020.
23. Q. K. U. D. Arshad, A. U. Kashif, and I. M. Quershi, "A review on the evolution of cellular technologies," In *2019 16th International Bhurban Conference on Applied Sciences and Technology (IBCAST)*, 2019: IEEE, pp. 989–993.
24. F. Al-Turjman and A. Malekloo, "Smart parking in IoT-enabled cities: A survey," *Sustainable Cities and Society*, vol. 49, p. 101608, 2019.

25. M. Khir et al., "Framework design of traffic light offense camera monitoring surveillance system," In *2018 International Conference on Intelligent and Advanced System (ICIAS)*, 2018: IEEE, pp. 1–4.
26. S. Kok, A. Abdullah, N. JhanJhi, and M. Supramaniam, "Prevention of crypto-ransomware using a pre-encryption detection algorithm," *Computers*, vol. 8, no. 4, p. 79, 2019.
27. S. Kok, A. Abdullah, N. Jhanjhi, and M. Supramaniam, "Ransomware, threat and detection techniques: A review," *International Journal Computer Science and Network Security*, vol. 19, no. 2, p. 136, 2019.
28. S. Hajiheidari, K. Wakil, M. Badri, and N. J. Navimipour, "Intrusion detection systems in the Internet of things: A comprehensive investigation," *Computer Networks*, vol. 160, pp. 165–191, 2019.
29. T. Soyata, H. Habibzadeh, C. Ekenna, B. Nussbaum, and J. Lozano, "Smart city in crisis: Technology and policy concerns," *Sustainable Cities and Society*, vol. 50, p. 101566, 2019.
30. M. Humayun, N. Z. Mahmood Niazi M. A. Jhanjhi, and S. Mahmood. "Cyber security threats and vulnerabilities: A systematic mapping study," *Arabian Journal for Science and Engineering*, vol. 45, no. 4, 3171–3189, 2020.

10

A Theoretical Framework on Enhancing Cloud Storage Security through Customized ECC Key Management Technique

G. Prakash
Amrita School of Engineering, Amrita Vishwa Vidyapeetham, Bengaluru, India

M. Kiruthigga
Redfowl Infotech Pvt. Ltd., Namakkal, India

CONTENTS

10.1 Introduction

Cloud computing technology is currently transforming the way the world does business. Cloud computing is a fundamental technology that allows for the sharing of resources over the Internet. A majority of the organizations are shifting toward contemporary cloud computing technology to maintain their

DOI: 10.1201/9781003203087-10

large volume of data backups, enhanced collaborations, scalable easy access, and infinite data storage space with low operational cost. Cloud computing will make it possible to consume services on demand. Cloud computing features include on-demand self-provisioning, ubiquitous network elasticity and connectivity, location-independent resource pooling, quick resource flexibility, consumption-based pricing, and risk transfer to a third party. The benefits of outsourcing cloud technology for bulk data storage have attracted the attention of both the industrial and academic research communities.

Cloud computing is very promising for IT applications; nevertheless, there are still some issues that need to be addressed. Doshi, Riddhi, and Vivek Kute [1] introduced Cloud computing enabling the users to generate, personalize, and set up online applications. It allows the users to access software utilities and a pack of resources adaptively. It is a computational methodology where a large number of system nodes are linked together privately or publicly to deliver continuous, highly available network stacks for various data file storage. It offers constrained consumer-operated service accesses and utilization of data resources over the Internet. Even though it offers significant services, it deals with security issues. Many potential cloud applications suffer from the generic approach's concern, in which a shared cloud service becomes unstable and insecure as a result of commercial demands from IT and other corporate organizations. As "resource pooling" is considered to be one of the most important characteristics, the same feature becomes critical when it was shared with other cloud users. As a consequence of data leaking, unstable data interface and destruction of data have led to larger risks and security issues.

While deploying data applications in the cloud eco-system, the significant issue is that of maintaining privacy during sensitive transactions. The challenging factor is how data can be transmitted with assured privacy where many cloud services are outsourced to multiple users, especially with the Storage-as-a-service. Of course, privacy concerns do not influence all models to the same degree pertaining to the cloud deployment models. When compared with the public and hybrid cloud models, the private cloud is less affected. In fact, the former models are open (more or less) to the general public (individual, business, etc.), whereas private cloud resources are tailored exclusively for a single entity (e.g., an enterprise). It is the job of this organization to manage and regulate these resources. The term "private cloud" refers to a personally identifiable information system that adheres to the cloud paradigm, including its features, characteristics, and service tiers.

Malicious assaults on user data stored in cloud systems are common for any business entity, with the primary goal being obtaining user privacy before obtaining economic gains. In this case, the most pressing issues to be resolved are conventional laws, regulations, and processes, where these existing legal policies are not adequate for securing the business-related sensitive data. Hence, relevant laws and regulations should be established and improved to provide third-party systems with trustable and privacy-preserving security, to meet company requirements. To provide clear responsibility partition when problems arise, as well as to deliver protection mechanisms when cloud service providers' (CSPs) exit should also be considered. Most domestic entities are wary of entrusting sensitive data to third parties, preferring instead to establish their own private cloud. When data are taken outside the company, Data Privacy settings are beyond the control of the enterprise. Several risks are associated with the use of third-party cloud services. One significant risk is **Key Management** where the Management of cryptographic keys takes on an additional complexity in cloud environments.

Data privacy can be simply given by encrypting data before it leaves the organization's perimeter, either with symmetric/asymmetric encryption or with current cryptography such as Identity-Based Encryption (IBE) or Attribute-Based Encryption (ABE). IBE is the process of an associated private key that is extracted from the public key by an authorized third-party server. It is, nevertheless, the organization's responsibility to manage and send the cryptographic keys securely through the third-party cloud services. Using client-side cryptography improves data secrecy; however, most proposed solutions focus on data security rather than key security, making key confidentiality less secure, as demonstrated by key leakage during data transfer in FADE's architecture. Modern encryptions, such as IBE and ABE, do not indicate that Secure Key Exchange should be given more attention.

The rest of the chapter encompasses the following: Section 10.2 discusses the concerns in managing the Secure Key Establishment while providing the cloud storage to consumers through the Infrastructure-as-a-Service (IaaS). Section 10.3 details the background works to demonstrate cryptographic key management techniques to be analyzed in accordance with various methodologies. Section 10.4 details the proposed theoretical framework and key management model. The Mathematical Modeling using Elliptic Curve Cryptography (ECC) key generation is discussed in Section 10.5. The theoretical analysis regarding the strong resistance to attacks on using ECC-based Key Management in Cloud Storage systems is detailed in Sections 10.5 and 10.6. Section 10.7 concludes the proposed theoretical work and the need for real-time implementation in the near future.

10.2 IaaS Key Management Concerns

IaaS refers to on-demand access to cloud-hosted infrastructures, such as servers, storage, and networking resources, that the cloud users may install, configure, and utilize in the same manner they do on-premises hardware. The distinction is that the CSP hosts and manages the hardware and storage resources through their own cloud data centers. Customers who utilize IaaS, pay on a subscription or pay-as-you-go basis for the storage systems they use through the Internet itself. Security threats and requirements vary widely depending on the deployment model (Hybrid, Public, or Private) in the IaaS environment. Millions of variables and technologies influence the list of IaaS security threats, especially on sensitive data transactions through cloud storage databases.

Ahmed Bentajer and Mustapha Hedabou [2] in their paper highlighted two techniques for managing the cryptographic keys by the CSP. One is the Bring Your Own Key (BYOK) and another is Key Management Interoperability Protocol (KMIP). BYOK was first offered by Amazon to allow enterprises to import their own Key Management Infrastructure (KMI) into the Amazon Web Services (AWS) Key Management Service (KMS). This feature gives the Service Providers more control over key management and allows them to keep the Cloud Customer/User keys safely within the Hardware Module.

Amazon proposed AWS CloudHSM for this purpose, which runs on a virtual private cloud under the authority of the business. Hardware Module being physically located in the Cloud data center allows for the creation of its own master key, which is used to encrypt data. Because it is hidden in the Hardware Module, no one can read it, not even the Cloud Hardware Module provider. Concerns about key types and strengths (AES256 vs. AES-128 or 3DES), as well as key generation sources (Hardware Module versus Software), can be addressed with BYOK. As a result, BYOK can be useful in any business auditing process that deploys Cloud Infrastructure. When the keys are posted into the CSP KMS, however, customers are not having complete control over and exclusive access to those cryptographic keys [3].

Haikun Liu et al. [4] worked on an efficient, novel Hemera Virtual Machine Image Management to provide scalable services for cloud consumers with controllable vulnerabilities scanning/auditing facilities. But this approach is structured for database operations and focused toward the performance improvements of VMI operations. In their work, the authors have not specified the Secure Key Management techniques between the Cloud consumers and VMI supported database and hence it is a debatable consideration of how the consumers could have secure access and control of their data stored in the cloud.

10.3 Related Work

Heli Amarasinghe et al. [5] implemented a unique virtual network management system based on protection and restoration to improve fault tolerance in IaaS and optimize IaaS resource allocation on SDN-based cloud infrastructure. Multiple geographically disseminated data centers are interconnected with SDN to form the networked cloud infrastructure. To assess the protection and restoration methods, the authors ran a series of tests. The findings show that their concept outperforms existing protection measures and can improve IaaS request acceptance and lower surface resource consumption.

Ansar Rafique et al. [6] proposed a new CryptDICE that hides the implementation and management complexity of various data encryption algorithms in NoSQL database applications. The lightweight service provided by CryptDICE simplifies the process of developing User Defined Functions (UDF) directly in the database engine for each underlying database technology. As a result, the service uses UDF in its application code and provides migration transparency, allowing it to be moved from on-premises to public clouds and perform complex computations alongside the database engine to deliver low-latency aggregate queries. The current preliminary assessment is largely concerned with functional features, performance standards, and latency factors. The authors developed a secure layer for data access where an Encryption API was present to encrypt the entire NoSQL queries using the AES algorithm. However, there were no specific details on how the query processor has shared the secret key between the processes present in the Query Execution System.

Yongkai Fan et al. [7] were claiming that the integrity of data becomes a possible vulnerability when it is delivered to an unknown CSP. The authors proposed a new data integrity checking scheme called Secure Identity-Based Aggregate Signatures (SIBAS), which uses a Trusted Execution Environment (TEE) to replace a third-party auditor with a single secure environment on the client side that securely checks the integrity of outsourced data without relying on cloud meta-data. Through Shamir's (t,n) threshold scheme, SIBAS may not only validate the integrity of outsourced data, but also secure key management in TEE. It is to our advantage in this strategy to preserve the confidentiality of secret keys. The authors' assumption focused on mitigating the attacks highlighted by Diffie–Hellman Key Exchange and created the Private Keys using the Trusted Third Party which may not be suitable in the cloud storage environment.

It is required to keep the cloud users authenticated and not violate the security features while the data was transferred through the cloud service. The Multi-Factor Authentication (MFA) with Public Key Infrastructure (PKI) for authenticity and various security policies required for end users in the cloud services has been employed to prevent cyber-attacks on the cloud computing which was explained by Priteshkumar Prajapati and Parth Shah [8].

The user verification in cloud computing is the crucial step aimed at data security. It is considered that the key exchange and process of encryption are complex to achieve in a complicated context. Therefore, image-based authentication is the technique which has been employed in secured key exchanges and can work based on the sequence of selected images. Tomar A.S. et al. [9] proposed an additional mechanism of including ECC technique along with the Captcha to detect machine users. Since it is enabled with a small size key with fewer computation steps, it renders secure authenticated transactions in the cloud storage systems.

Furthermore, the problems in sending data to unauthorized service providers of cloud services have led to potential consequences on database security. Fan, Yongkai et al. [7] provided a solution for this problem such that it can be decrypted by SIBAS since the data protection is enabled through the verification strategy of TEE. The TEE can act as an inspector which guarantees the data secure environment.

Correspondingly, the data masking approach developed by Wang, Liang et al. [10] can potentially reduce the data duplication in the cloud database by storing one copy of the redundant data instead of multiple copies. Even though the received redundant data have been employed with Convergent Encryption (CE) strategy, this approach tackles the problems of conceptually not secured CE and sensitivity with offline brute-force attacks. Another problem due to Convergent Key (CK) management has brought up the additional pressure in third party generation due to the employment of the same keys for redundant duplicates. Thus, the repetition of several typical keys' storage raises key management issues. Hence, the key-sharing method with safe deduplication can address the issues, which can be done with the working operation of randomly chosen convergent keys, so the testified data can reclaim the CK.

In a cloud virtual infrastructure, the requirement for data confidentiality has been raised due to the weakened third parties. Accordingly, data protection is one of the significant factors to be examined while exchanging the data. Velumadhava Rao, R. et al. [11], provided a secure mechanism where the master password or private key is enabled and the complete file is encrypted effectively in a traditional architecture. Although this private key mechanism is not applied to the cloud-based user groups owing to the key distribution. To be operated under group key methodology for a cloud-based architecture, a hierarchical group key management is employed for protected data exchange on the cloud infrastructure. This method has been performed by enabling Key Distribution Server (KDS) for protecting cloud data by enabling group keys based on the KDS secret value. The Logical Key Hierarchy (LKH) protocol is used in this technique to establish a hierarchical tree for the adaptivity of data exchange.

Hosam, Osama, and Muhammad Hammad Ahmad [12] studied and analyzed that evolving cloud technologies have a profound effect on data processing, retention, confidentiality, and efficiency. As cloud computing systems reduce the effect of bulk transmission errors and result in positive outcomes, managing challenges with certain levels, the rapid storage of data in a cloud system brings radical changes in the business organizations that processes huge data. The real-time customizable services deployed through cloud service technologies produce an ideal alternative for consumers and businesses across the globe. The significant security vulnerabilities in the cloud should always be resolved by encrypting the confidential information and preserving the privacy of a large number of cloud users. To fix the challenges of key generation, the hybrid cloud provides the data security enabled with the various algorithms such as strong symmetric Advanced Encryption Standard (AES), asymmetric ECC, and the Least Significant Bit (LSB)-based cryptanalysis technique in the cloud. Unless the customer chooses to share the sensitive information with the second user, the system won't implant the AES key. So, the customer who requires to send the secure data should embed the AES key in the second user's picture. Hence, a strong cyber resilience and access control resource management for various users using the modern encryption algorithms ECC, and AES were achieved.

Xu, Cheng et al. [13] explained that according to the cloud-based streaming media systems, it can experience numerous complications owing to their client workstations with heterogeneous methodological requirements. A complex network environment and individualized user expectations give rise to new contradictions to cloud streaming media. The multilayer encryption algorithm for Scalable Video Coding (SVC) bitstream has been employed which was bitstream destined. The bitstream was segregated and encrypted as per the multilayered bit programming code of SVC by reorganizing the network abstraction layer (NAL) unit of the SVC bitstream. The multilayer feature of SVC is therefore protected in a hierarchical manner where multiple cryptographic algorithms for such base layers with enhancements can be applied subsequently in order to enable adequate protection and also augment computational effectiveness.

Kanna, G. Prabu, and V. Vasudevan [14] deployed the Fully Homomorphic–Elliptic Curve Cryptography (FH-ECC) technique to provide an essential privacy protection mechanism that has been employed to increase security protection. While putting the data on the cloud, the data owner encrypts the original data by converting it to encrypted format using the ECC algorithm and performing FH operations on the encrypted data. Once the user transmits the data packet to the cloud, the CSP evaluates the user's access control mechanism to see whether constrained access to the system is authorized. If the access control policy is credible, the user receives secret messages from which the encrypted data are retrieved. The ECC decryption and FH procedures are then used to retrieve the original information.

Liang Tan et al. [15] implemented a blockchain-based Cyber-Physical Social System (CPSS) big data that could solve the access control concerns with respect to privacy-preserving of bulk data stored in the cloud. For their approach, the authors processed the authorization, auditing, and access control security mechanisms through the use of a lightweight symmetric encryption algorithm. In spite of the performance issues such as

Computation Overhead, Storage Overhead, and Throughput in their system, as the encryption algorithm used was symmetric, the effective management of key establishment that has to be shared between the different cloud users was not clearly defined.

Subsequently, an innovative hybrid strategy is introduced to secure data communications when it has to be sent through cloud services, as well as a distinct digital signature is produced at each use side and recorded on a distributed set of blocks. To assess the hypothesized architecture, an online cloud similar to the actual cloud service system has been deployed. Despite the increasing computing resources taken to accomplish the conceptual methodology owing to the blockchain interface, the study found that information security and trustworthiness are retained, and privacy information is improved. In defiance of many benefits afforded by the cloud, considerable privacy and security issues have been raised, including privacy vulnerability, information leakage, tampering, Service-Level Agreement (SLA) breaking, and issues surrounding hybrid deployment patterns as identified by Marwan Adnan Darwish et al. [16]

In addition to the many researches done as yet on securing and managing the secret keys in cloud storage systems, this section also includes another work done by A. Kumari et al. [17] which proclaims a safe and efficient technique in terms of computing and communication costs. A key establishment architecture based on smartcards and ECC has been employed. This research implementation would also give real-world applications in the cloud and similar environments.

D. ManJiang et al. [18] dealt with the issue of inadequate malware protection caused by typical cryptographic protocols in power bidding system cloud data storage and have proposed a hybrid encryption solution based on both symmetric and asymmetric encryption. In this method, the bulk plain text is first partitioned into multiple blocks based on the quantity, and then the larger plain text block is encrypted using the notorious viable symmetric encryption algorithm AES, and the smaller data block along with the AES key are again encrypted using the asymmetric encryption algorithm ECC to verify the security resilience of plain text and privacy of the key details.

Syam Kumar P. and Subramanian R. [19] proposed an ECC and Sobol sequence-based protocol that is both efficient and secure. The proposed method is best suited for lean users with few resources and limited computing power. It meets all of the cloud data storage's security and performance requirements. Their method also provides public provability, which allows Third-Party Auditors to verify data integrity without retrieving original data from the server and detect data malpractices with high probability. Furthermore, this scheme supports dynamic data operations, which are performed by the user on data stored in the cloud while maintaining the same level of security. Through security analysis, the authors demonstrated that the proposed scheme is secure in terms of integrity and confidentiality.

Determining the security of advanced computational systems assembled together, as W. Jansen and T. Grance [20] explained in their NIST paper, is also a long-standing security challenge that affects large-scale computing in general, and cloud storage in specific. Attaining high assurance characteristics in system implementations has long been an unattainable objective for information security researchers and practitioners, and it's still a work in progress for cloud computing, as shown in the examples in this research. However, public cloud computing is a persuasive computing paradigm that government organizations should investigate for their IT solution set.

The challenges connected to cloud data storage noted by Nareshvurukonda and B. Thirumala Rao [21] include data breaches, data theft, and cloud data unreliability. Finally, they offered potential cloud-based solutions to various problems.

Salim Ali Abbas and AmalAbdulBaqiMaryoosh [22] developed a more agile and efficient solution to overcome the data storage security issues in cloud computing. The usage of Identified-Based Cryptography (IBC) allows for key management without the need for a certificate to be provided. Data security and confidentiality are protected by ECC before being stored on cloud servers, which also use the Elliptic Curve Digital Signature technique to ensure data integrity.

Tara Alimunisha and Kolli Nuka Raju [23] proposed a computing model for big data technology on the cloud and also did a survey of many cryptographic algorithms that can be deployed to secure these analytics in a variety of aspects. Owners of big data must consider the complexity of encryption (both in terms of time and money) due to its massive size. Their proposed method avoids this by distributing data across multiple cloud providers and encrypting it with the ECC algorithm. Based on a conditional proxy re-encryption technique, the proposed methodology is a content sharing scheme that is safe in cloud storage systems. Related to various characteristics, this method considerably reduced a client's load. The first is that the re-encryption process is assigned to a cloud server, with the client's involvement limited to data encryption and decryption, as well as the creation of re-encryption keys. Second, the number of re-encryption keys that must be shared is kept to a minimum.

Because SSL/TLS is the most extensively used security protocol on the web, more efficient SSL/TLS will have a substantial effect on Internet performance, according to Benjamin Clement Sebastian and Ugur Alpay [24]. Their article evaluates the two cryptosystems to see if ECC has a significant performance improvement over Rivest–Shamir–Adleman (RSA) when used in the SSL/TLS protocol.

M. Gobi and Karthik Sundararaj [25] suggested an ECC-based methodology to improve data security in cloud computing. The difficulty of computing discrete logarithms in a finite field is used to improve and maintain security. El-Gamal and ECC are two types of public-key cryptography in which one decryption key, the private key, is kept hidden while the other, the public key, is freely shared.

10.4 Problems Identified

Many companies are transitioning some or all of their activities to the cloud, which has necessitated the transition of their security. The process of keeping secure track of managing cryptographic keys within a cryptosystem is one major critical task that has to be considered. The good news is that a number of key management companies have teamed up with cloud hosting companies to deploy traditional Hardware Security Modules in cloud environments. As it is a standard Hardware Security Module in an offsite environment, the same levels of "hardening" would be required. The flexibility of virtual encryption key managers is far greater than that of Hardware Security Module counterparts. In many circumstances, a virtual key manager may be acquired from a vendor and installed in a virtual environment in a matter of minutes. A Hardware Security Module, on the other hand, can take days or weeks to arrive at its destination.

Virtual instances can also be placed anywhere that supports the virtual platform used by the key manager, such as VMware. The drawback, of course, is that a virtual key manager's software can only be Federal Information Processing Standards (FIPS) 140-2 compliant, not verified, due to the fact that it is virtual and has no fixed physical components. A Hardware Security Module is the only option if the company demands or compliance regulations necessitate FIPS 140-2 validation. However, the logical security provided by FIPS 140-2 compliant virtual key management is usually sufficient for most organizational purposes. Encryption key management is available through marketplace products from cloud providers such as AWS and Microsoft Azure.

Cloud users must preserve the private key of the public/private key pair that they employ to authenticate themselves using enterprise security procedures. The Diffie–Hellman keys, as well as the session keys derived from them, are ephemeral and generated or computed on the fly. As a result, these keys don't need to be stored for a long time, and key management isn't a problem. In most infrastructure cloud services, the customer either deploys its own virtual machine-based computing resources or leases them from the CSP. The pre-images of the Virtual Machine are authenticated using keys that are only known by authorized sources and should not be tampered with. The cloud Consumer must subscribe to a Database in order to store structured data generated by applications operating on the Virtual Machine.

10.5 Case Study Analysis – Amazon Web Services Key Management Service

10.5.1 Centralized Key Management

To proceed further with the Key Management concerns in the IaaS Cloud, a case study on AWS is studied and analyzed. AWS© KMS offers the centralized control of maintaining and accessing the public and private keys. KMS is managing these keys and provides them to the concerned users when required. AWS Customers can produce new master keys at any time and govern who has access to them and which services they can get them with. These keys can also be imported, in addition to newer keys being created as needed by the cloud user. Such keys are stored in the AWS© Cloud Hardware Security Module cluster. Master Keys in KMS (imported or produced on behalf of customers) are encrypted and kept in a highly durable form to enable and ensure that they can be retrieved when needed. To secure the confidentiality and integrity of their customer keys, AWS uses FIPS 140-2 verified cloud hardware security modules (CHSMs). Customers' plaintext keys are never written to disks and are only stored in the CHSMs' volatile memory for the period of the desired cryptographic operation. KMS keys are never transported outside of the AWS zone, as seen in Figure 10.1 in which they were created.

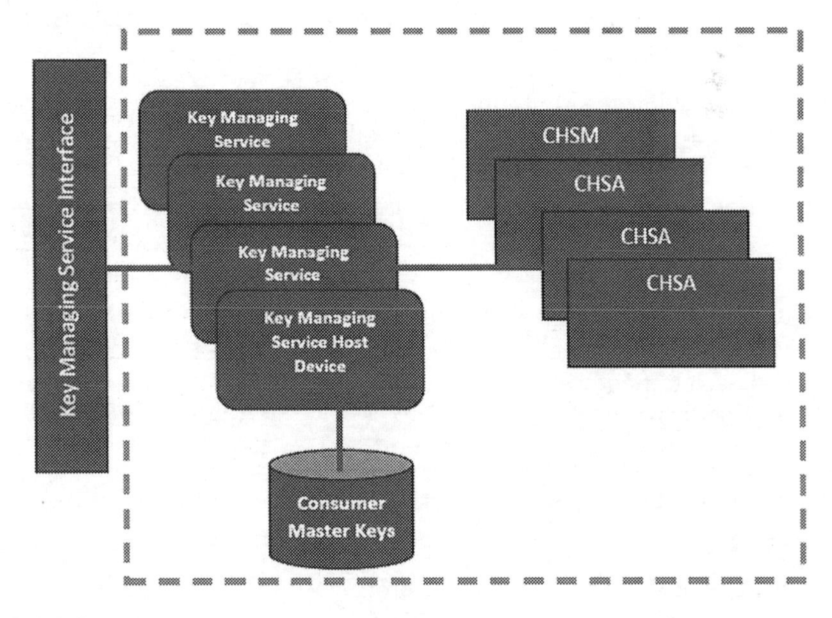

FIGURE 10.1
Amazon Web Services © Key Management Service (AWS KMS).

In AWS, to implement structured data applications, IaaS Cloud consumers are required to subscribe to a database management and storage system. It has been found that many enterprise consumers are now migrating their existing databases to AWS without enhancing the features such as scalability, cost-effectiveness, and capability to cope with the native cloud services. This kind of direct migration would enable superfluous operational overheads and additional complexity leading to substantial time for responding to events. Transparent Data Encryption (TDE) is the database technology used by AWS for encrypting and decrypting the cloud-stored data which are present in the MS-SQL Server and Oracle Enterprise Database. But the major problems that could happen in implementing the native encryption function are discussed and listed as follows:

(a) A single Database Encryption Key (DEK) protects the entire database.
(b) Because encryption occurs at the I/O level, the DEK must be located near the storage resources dedicated to database storage space.
(c) The only option for the Cloud Consumer is to store the DEK in the same cloud as the DBMS instance.
(d) Difficulty of assigning various encryption keys to different users depending on their permissions.

In addition to DEK, users can encrypt data at the column, table, or even a set of data files corresponding to several tables or indexes, which is referred to as User level encryption (ULE). However, the following issues arise as a result of this ULE:

(a) Make a call to a KMS to map the user's set of session permissions (depending on the roles assumed) to the set of keys, for retrieving the required set of keys from key storage.
(b) During a user's session (for key usage), the keys are cached in the same cloud as the DBMS instance in a memory space established for the user session.
(c) There's also the issue of collecting the key from the KMS and securely delivering it to the application executing in the CSP environment.
(d) Unless additional security measures are adopted, this technique makes sensitive data exposed to access by CSP Administrators.

Hence, to address the above concerns, the following specific concepts are planned for the preliminary work to be done as part of the theoretical framework:

- To establish cloud security access broker (CASB) solutions that really can guarantee and enforce specific data privacy and security standards for both storage and processing.

- To provide significant security for accessing cloud storage systems through a strong ECC algorithm.
- To propose an additional complexity for both CSPs and cloud users through management of cryptographic keys in cloud environments with Elliptic Curve Diffie–Hellman Key Exchange Algorithm.

10.6 Proposed Theoretical Framework

As discussed in Section 10.5 for addressing the Key Management issues in Cloud Storage systems, the first concept which has to be considered is to identify the CASB for providing adequate data security and privacy. It is a cloud host software which is easily transitioned between the cloud consumers and CSPs. The CASB's functionality is to address the concerns in the IaaS by introducing the security policies that can be deployed while moving from the desktop operations to the cloud storage operations within the cloud-specific context. This transition is applicable for all such sensitive data that has to be protected from malicious users, as cloud infrastructure is normally a third-party storage system.

It has been found that many cryptographic primitives are used in the AWS KMS. These primitives provide the essential security and are easily configurable for the cloud systems. The default set of algorithms approved are AES in Galois Counter Mode for encrypting the cloud data; Cryptographically Secure Pseudo-Random Number Generator (CSPRNG) for generating the seed data; Elliptic Curve Digital Signature Algorithm (ECDSA) for authenticating between the cloud users besides key establishment.

According to AWS, two different Key management techniques are used: **First: A (1, 2, ECC DH) where 1 – Ephemeral Key, 2 – Static Keys**. This scheme has a creator, and a static key used for signing. The creator designates his signature as an ephemeral Elliptic Curve Cryptography Diffie–Hellman (ECCDH) key, proposed for the receiver with a second static ECCDH agreement key; **Second: B (2, 2, ECC CDH) where 2 – Ephemeral Keys, 2 – Static Keys**. This scheme is designed such that both parties have a static signing key and they write a signature and exchange an Ephemeral ECCDH Key between themselves. It is critical that the Diffie–Hellman keys, as well as the derived session keys, are ephemeral and produced or calculated on the fly. As a result, these keys do not entail long-term storage, and hence, the key agreement is not a concern. However, the problem arises due to the nature of using static keys for secure transactions in the Cloud Storage systems.

10.6.1 Static Key Usage in Symmetric and Asymmetric Encryptions

A Cryptographic Static key is one that has been used over a long duration in a secure transaction. It is frequently used and considered less secure than using an ephemeral key. This is due to the reasons that an eavesdropper could have more information to work with. If the key is broken, an attacker can decrypt several messages. Also, it allows the attacker to initiate a known plaintext attack. However, Static keys are easier to install and hence it has been used in many Cloud storage services. The provision of security response is that "it depends, sometimes strong, sometime weak". If the static key is part of the cloud application, then an attacker can easily extract it by using any Reverse Engineering tool. It is not required to break any encryption at all.

Static Key (Personal Code) Cryptography and Heterogeneous Key (Common Code) Cryptography are the two different forms of Cryptography. This has been reviewed by most of the standard algorithms based on Asymmetric Cryptography – which normally uses Key Exchange Encryption technology that employed static keys (the public keys) for both authentication and integrity services. On the other hand, Symmetric Key Encryption, also known as secret cryptosystem, represents the strategy of setting up a secret key, the only single key used by the cryptography algorithm shared between the communicating parties. In such terms, authentication processes are done by default by using that single key only for initiating the integrity of communicating parties. The most well-known algorithms that employ this strategy are Data Encryption Standard (DES) and AES algorithms.

As a result, it significantly outperformed its Asymmetric Encryption equivalent due to the implicit authentication and integrity measures available as a secret key shared only between the concerned authorities. However, there will be another issue of how this secret key will be shared securely, especially in the cloud storage systems. As the cloud aims for advanced technologies in terms of providing Software, Infrastructure, and Platform as the services for various authorized cloud consumers, it requires making use of effective secure key establishments. Due to this purpose, Cloud Storage systems should organize and utilize the Symmetric and Asymmetric Encryption algorithms based on how the dynamic keys outperform the static keys. The cryptographic details pertaining to the downsides and upsides of using static keys and dynamic keys are represented in the Tables 10.1 and 10.2 clearly as follows

Based on the above tables, it is clear that in a real-world scenario, how the public and private keys are used in the various encryption algorithms and their practical state of either "static" or "dynamic" determine the secure

TABLE 10.1

Methodologies with Static Key

S. No.	Approaches	Functionalities	Downsides	Performance Parameters
1.	**Symmetric key encryption algorithm based on cyclic elliptic curve and chaotic system**	It consists of a set of huge digits that are generated using some approaches to authenticate data. Secret keys are used to produce signatures, which are then confirmed using public keys.	Distinct Testing software is required to check the integrity. Digital certificates should be obtained from authorized entities.	It is extremely quick, and it ensures non-repudiation and authentication. It is more beneficial for peer-to-peer systems.
2.	**NJJSAA symmetric key algorithm**	For both encrypting and decrypting, the methodology employs shared keys and Opcode procedures.	The approach is a time-consuming and prone to probable timing attacks.	It outperforms other overall cryptography methods. It is capable of encrypting both big and small files.
3.	**Symmetric key-based RFID authentication protocol**	It adopts identical RF-based tools to develop multiple protocols that apply block cipher modes of operation.	The procedure is also long and time-consuming.	This protocol enhances the RFID authentication mechanism by offering protection over various threats while requiring less computing effort.
4.	**DJMNA symmetric key algorithm**	It incorporates the MGVC and DJSA techniques. The sequence with these procedures is evaluated by the randomized matrix generated through the operation.	The method is tedious. The methodology that is followed is complicated and time-consuming.	The encryption process is extremely difficult to decipher using any brute-force attack.

TABLE 10.2

Methodologies with Dynamic Key

S. No.	Approaches	Functionalities	Upsides	Performance Parameters
1.	**Elliptical Curve Cryptography**	It generates the key code using elliptic curves. Actually, accomplished with a 160-bit key code.	It is safer and stronger than the RSA and Diffie–Hellman algorithms. It requires less power and enables reduced energy consumption.	The Elliptic Curve Digital Signature Algorithm (ECDSA) is developed. Elliptic Curve Menezes-Qu-Vanstone (ECMQV) or Authenticated Key Exchange can be generated for cloud systems.
2.	**Diffie–Hellman Key Exchange**	It is entirely focused on the exchange of a public cryptographic key. This key is required for encrypted communications as well as decryption. It is primarily focused on the hard problem involved in discrete logarithms.	Since the secret key is so small (256 bits), the approach is relatively fast, but prone to Man-in-the-Middle Attack	It is necessary to shift keys on a routine basis. Attacks are thwarted in the worst-case scenario by developing a tunneling protocol. A response to the attacks is also the creation of digital signatures.
3.	**Rivest–Shamir–Adleman (RSA)**	The basic form is (d, e), in which d is the secret key and e is the public identifier. Relative operation is implemented in both encryption and decryption.	It is hard to generate the secret key from the master password and modulus, making it increasingly protected. The hackers have such a great difficulty estimating the inverse of e.	The size should be greater than 1024 bits.

(*Continued*)

TABLE 10.2 (*Continued*)

Methodologies with Dynamic Key

S. No.	Approaches	Functionalities	Upsides	Performance Parameters
4.	**Digital Signature Algorithm (DSA)**	It consists of a set of variable integers given as input that generates fixed secret code to authenticate data. Secret keys are used to produce signatures, which are then confirmed using public keys.	It is extremely fast, and it ensures non-repudiation and authentication. Highly commercialized in many web applications.	Distinct Testing software are required to check the integrity.

operations. Hence, it is now mandatory to find out the essential security requirements for secure key distributions which are listed as follows:

(i) It is not required to register with any kind of CSPs.
(ii) Authentication is to be done mutually between the cloud users and the services only.
(iii) Secure sessions should be created only during the initiation of key transfers and agreements between the cloud clients and servers. At any particular moment, even the CSPs are not supposed to obtain any information related to the secret key used.
(iv) The secret key of the cloud users and their data should ensure anonymity and privacy such that no external entities can obtain the cloud customers' actual keys and their identity while performing the authentication.
(v) To monitor the attackers' or intruders' identity, the proposed model should guarantee that there is just one identity capable of revealing the cloud customer's true identity.
(vi) To safeguard previously transmitted messages, the protocol should ensure that
    (a) any attacker, even if it acquires the private key of cloud users, is unable to reconstruct the earlier session keys
    (b) The proposed model should be resilient against other generic attacks.

When cryptographic keys are used to secure sensitive information, they must be carefully handled. To ensure that keys are kept safe, sophisticated key management systems are often utilized. The keys are:

(i) Created to the desired length using a well-protected high-quality random data source normally using the Hardware Security Module
(ii) Only authorized individuals are responsible for managing the security where the policies are strictly adhered to
(iii) Utilized just for the functions they were designed for
  (a) Updated according to their crypto-duration, and discarded when no longer needed;
  (b) Fully auditable to offer proof of correct (or erroneous) usage. Other attributes that enable keys to be modified and controlled according to pre-specified policies are frequently defined by key management systems.

10.6.2 AWS Cryptographic Key Establishment

It has been already discussed that AWS utilizes various cryptographic services which will assure the integrity of cloud data at rest or transition. The AWS KMS provides CloudHSMs which ensure storage of different keys backed by FIPS 140-2 standard. CloudHSMs are based on the operation of multi-Tenant architecture that could generate and manage these keys. Also, by integrating CloudHSM and AWS, the encryption keys will be available for synchronization, and backup, and could easily replace the failure modules, thus making the key distribution manageable. But there will be other overheads of user privileges, scalable issues, and logs that have to be taken care.

Generally, AWS KMS Key storage is having its limitation with regard to the compliance of FIPS 140-2 Hardware Security Module. The responsibility related to performance, auditing, and user management is now shifted to the cloud consumers. At this point, the user has to clearly identify the security vulnerabilities on the Key Management and find out an optimum solution to mitigate the threats initiated by the Cloud attackers. While considering AWS KMS service as adequate for any Cloud Storage Infrastructure, the requirements on compliance FIPS 140-2 Level 3 lead to a specific class of vulnerability referred to as DUHK attack. There are numerous vulnerable implementations while examining the Information Security policy documents from 288 vendors who used the FIPS 140-2 certified keys. It was stated by the concerned researchers that Hard-Coded Keys are prone to the vulnerabilities that exist in the usage of ANSI X9.31 Random Number Generator (RNG). This algorithm is commonly used in most web applications for generating the cryptographic keys to prevent third parties from accessing interrupted transactions.

It is well known from Section 10.6.1 that AWS KMS uses the AES algorithm along with CSPRNG for their cloud data encryption. But due to the DUHK attack, the FIPS compatible ANSI X9.31 PRNG is vulnerable. It has been analyzed by the authors who found this ridiculous attack that being a legacy software, ANSI X9.31 took two key inputs K and V, where K is the long-term key and V represents an initial state. The PRNG would produce a lengthy stream of pseudorandom bits by recurrently applying the data cipher block. Based on the functional aspects of ANSI X9.31 PRNG, the key K is never changed and remains fixed throughout the entire process which leads to a major problem that an attacker could easily grab the key K. If K is known, then the attacker could initiate the attack and find all the other ingredients through reverse engineering. Thus, if any cloud-based storage system uses this ANSI PRNG to produce the secret keys, an attacker could theoretically recover these keys and the entire encryption process would be broken. Hence, it is considered that AWS KMS, which is basically using this FIPS 140-2 validated Cloud HSM, is vulnerable to probable DUHK attacks. To overcome this problem, a proposed theoretical framework for Secure Key Establishment and Distribution has to be planned by improving the security features of the Elliptic Curve Diffie–Hellman Key Exchange Algorithm in the Cloud storage systems as shown in Figure 10.2.

CSPs offer a better database management infrastructure service for the cloud users where the database model includes appropriate security services. This database service model has a specialized Encryption Engine termed Transparent Data Encryption (TDE). A Database Management System

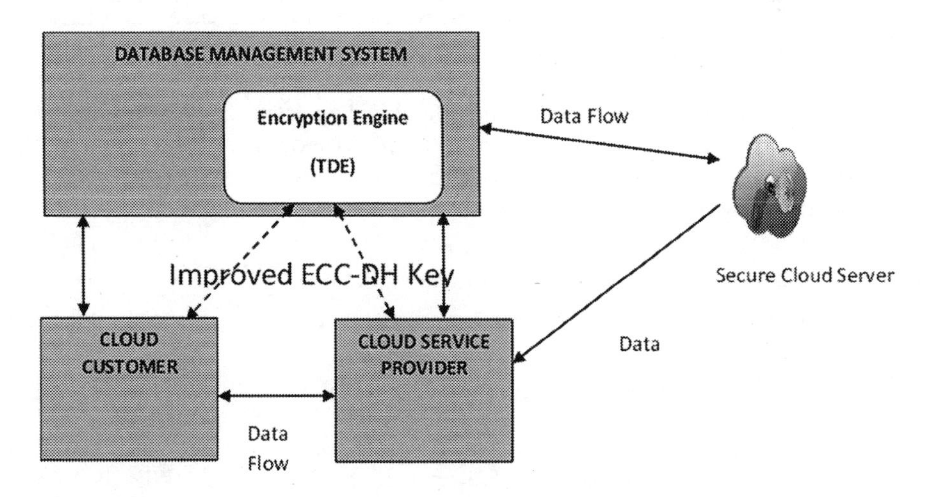

FIGURE 10.2
Proposed framework on secure key establishment and distribution.

(DBMS) instance is often offered to the cloud Consumer that subscribes to this service, with the opportunity to customize the instance to meet its business and security needs. The following stages outline the alternatives available for providing confidentiality protection for data handled by the DBMS instance, as well as the accompanying key management challenge:

1. TDE is a feature of the DBMS that uses a native encryption function.
2. The encryption engine works at the input–output level, encrypting data stored just before it is written to the external hard drive.
3. The ECC Diffie–Hellman Key Exchange encryption algorithm is used to safeguard the entire database.
4. There is no need to change the application logic or database schema because the TDE executes all cryptographic operations at the I/O level within the database system.
5. Because the cloud customer has administrative rights to the DBMS engine, it also has control over the ECC Diffie–Hellman (DH) Keys, which must be kept close to the storage resources designated for database data storage.
6. As a result, the cloud customer has no choice but to keep the DH Keys in the same cloud as the DBMS engine.

10.6.3 Customized Dynamic Key Management Service Using Blockchain-Based ECCDH Scheme

It has been proposed to have a blockchain-based key generation for the heterogeneous Cloud storage systems with restricted access control to authenticated cloud customers from their CSPs. Blockchain technology improvises dynamic keys while doing secure transactions through the public network which also provides lightweight and scalable key transfers for numerous cloud customers. This kind of blockchain structure deployed in cloud storage systems for secure key transfers will be decentralized based on its intrinsic functionality. Two major functionalities can be considered:

(i) Key Establishment through Blockchain Technology
(ii) Dynamic Key Distribution Scheme

A mathematical model can be constructed with ECC integrated Diffie–Hellman Key Exchange scheme for all the secure transactions of cloud customer's public keys that could run on cloud environment without any centralized access control. It is well known that the Decentralized Blockchain requires a client-side operation for a smart contract which holds different verifiable properties. This Smart Contract-based application could effectively make use of ECC-DH Key Exchange for cloud storage systems. Now, the key agreement between the Cloud users shall be verified and deployed without external trusted third parties. This blockchain integrated ECC-DH Key Exchange offers almost all the security services such as non-repudiation, integrity, and authenticity.

10.7 Conclusion and Future Work

As businesses become more reliant on cloud storage, data security is becoming a top responsibility in their IT architecture and information security initiatives. Companies are now recognizing the importance of protecting sensitive data while allowing employees to benefit from the cloud's performance and flexibility. Baseline security for cloud storage platforms and the data they process, such as authentication, access control, and encryption, should always be in place. Accordingly in this chapter, it has been identified that the cloud storage systems are vulnerable to stolen key attacks such that the intruders could get hold of the public/private keys during transit between the cloud customers and the service providers. This threat is due to the fact that the existing key establishment/management techniques are weak enough to not withstand the attacks. Hence, to access the keys securely, a customized and enhanced ECC Diffie–Hellman Key Management technique is proposed theoretically along with the blockchain technology and dynamic key distributions. From there on, it is essential that most enterprises in near future supplement these protections with added security measures of their own to strengthen cloud data protection and tighten access to sensitive information in the cloud.

References

1. Doshi, R. and Kute, V., "A review paper on security concerns in cloud computing and proposed security models," In *2020 International Conference on Emerging Trends in Information Technology and Engineering (ic-ETITE)*, pp. 1–4, 2020, doi: 10.1109/ic-ETITE47903.2020.37.
2. Bentajer, A. and Hedabou, M., "Cryptographic key management issues in cloud computing," *Advances in Engineering Research*, 151, 2020.
3. https://techbeacon.com/security/what-you-need-know-about-bring-your-own-key-cloud
4. Liu, H., He, B., Liao, X., and Jin, H., "Towards declarative and data-centric virtual machine image management in IaaS clouds," *IEEE Transactions on Cloud Computing*, 7, no. 4, 1124–1138, 2019, doi: 10.1109/TCC.2017.2728066.

5. Amarasinghe, H., Jarray, A., and Karmouch, A., "Survivable IaaS management with SDN" *IEEE Transactions on Cloud Computing*, 9, no. 4, 1619–1633, 2019. doi: 10.1109/TCC.2019.2910111.
6. Rafique, A., Van Landuyt, D., Beni, E. H., Lagaisse, B., and Joosen, W., CryptDICE: Distributed data protection system for secure cloud data storage and computation, *Information Systems*, 96, 101671, 2021, doi: 10.1016/j.is.2020.101671.
7. Fan, Y., Lin, X., Tan, G., Zhang, Y., Dong, W., and Lei, J., One secure data integrity verification scheme for cloud storage, *Future Generation Computer Systems*, 96, 376–385, 2019, doi: 10.1016/j.future.2019.01.054.
8. Prajapati, P. and Shah, P., "A review on secure data deduplication: Cloud storage security issue," *Journal of King Saud University - Computer and Information Sciences*, 2020, doi: 10.1016/j.jksuci.2020.10.021.
9. Tomar, A.S., Shankar, S.K., Sharma, M., and Bakshi, A., "Enhanced Image Based Authentication with Secure Key Exchange Mechanism Using ECC in Cloud," In Mueller, P., Thampi, S., Alam Bhuiyan, M., Ko, R., Doss, R., and Alcaraz Calero, J. (eds) *Security in Computing and Communications. SSCC 2016. Communications in Computer and Information Science*, vol. 625. Springer, Singapore, 2016, doi: 10.1007/978-981-10-2738-3_6.
10. Wang, L., Wang, B., Song, W., and Zhang, Z., "A key-sharing based secure deduplication scheme in cloud storage," *Information Science*, 504, 48–60, 2019.
11. Velumadhava Rao, R., Selvamani, K., Kanimozhi, S., and Kannan, A., "Hierarchical group key management for secure data sharing in a cloud-based environment," *Concurrency and Computation: Practice and Experience*, 31, no. 12, e4866, 2019.
12. Hosam, O. and Ahmad, M. H., "Hybrid design for cloud data security using combination of AES, ECC and LSB steganography," *International Journal of Computational Science and Engineering*, 19, no. 2, 153–161, 2019.
13. Xu, C., Ren, W., Linchen, Y., Zhu, T., and Raymond Choo, K.-K., "A hierarchical encryption and key management scheme for layered access control on H. 264/SVC bitstream in the internet of things," *IEEE Internet of Things Journal*, 7, no. 9, 8932–8942, 2020.
14. Kanna, G. P. and Vasudevan, V. "A fully homomorphic–elliptic curve cryptographybased encryption algorithm for ensuring the privacy preservation of the cloud data," *Cluster Computing*, 22, no. 4, 9561–9569, 2019.
15. Tan, L., Shi, N., Yang, C., and Keping, Y., "A blockchain-based access control framework for cyber-physical-social system big data," *IEEE Access*, 8, 77215–77226, 2020.
16. Darwish, M. A., Yafi, E., Al Ghamdi, M. A., and Almasri, A., "Decentralizing privacy implementation at cloud storage using blockchain-based hybrid algorithm," *Arabian Journal for Science and Engineering*, 1–10, 2020.
17. Kumari, A., Yahya Abbasi, M., and Alam, M., "A smartcard-based key agreement framework for cloud computing using ECC," In *2021 Third International Conference on Intelligent Communication Technologies and Virtual Mobile Networks (ICICV)*, pp. 43–48. IEEE, 2021.
18. ManJiang, D., Kai, C., Wang, Z. X., and LiPeng, Z., "Design of a cloud storage security encryption algorithm for power bidding system," In *2020 IEEE 4th Information Technology, Networking, Electronic and Automation Control Conference (ITNEC)*, vol. 1, pp. 1875–1879. IEEE, 2020.

19. Kumar, S. P., and Subramanian, R., "An efficient and secure protocol for ensuring data storage security in cloud computing," *International Journal of Computer Science Issues (IJCSI)*, 8, no. 6, 261, 2011.
20. Jansen, W. and Grance, T., *Guidelines on Security and Privacy in Public Cloud Computing (NIST SP 800-144)*, National Institute of Standards and Technology, U.S. Department of Commerce, 2011.
21. Rao, B. T., "A study on data storage security issues in cloud computing," *Procedia Computer Science*, 92, 128–135, 2016, doi: 10.1016/j.procs.2016.07.335.
22. Abbas, S. A. and Maryoosh, A. A., "Improving data storage security in cloud computing using elliptic curve cryptography," *IOSR Journal of Computer Engineering*, 17, no. 4, 48–53, 2015.
23. Alimunisha, T. and Raju, K. N., "Secure big data storage and sharing in cloud using ECC algorithm," *International Journal of Research*, 4, no. 5, 2017.
24. Sebastian, B. C. and Cenar, U. A., *Advantage of Using Elliptic Curve Cryptography in SSL/TLS*, Department of Computer Science University of California Santa Barbara, cs 290G fall term, 2015.
25. Gobi, M. and Sundararaj, K., "A secured cloud security using elliptic curve cryptography," *International Journal of Advanced Networking and Applications*, 12, 141–144, 2015.

Index

Printed in the United States
by Baker & Taylor Publisher Services